THE BRITISH POSTAL MUSEUM & ARCHIVE

Built:

POST OFFIC

Julian Osley

HINCKLEY
POST OFFICE
1902

Cover image: Plymouth post office, architect Cyril Pinfold, 1957.

Back cover image: North Finchley branch office, London, 1945, detail.

Title page image: Hinckley post office, architect William Thomas Oldrieve, 1902.

Contents page image: Keystone depicting Mercury, a recurrent motif for post offices, Hastings post office, architect David Dyke, 1927, detail.

First published 2010

Published by:
The British Postal Museum & Archive
Freeling House
Phoenix Place
London WC1X 0DL
www.postalheritage.org.uk

Text © Postal Heritage Trust 2010
This book is copyright under the Berne Convention.
All rights reserved.
ISBN 978-0-9553569-3-3

Designed by Crescent Lodge, London
Manufactured in the EU by Healeys Print Group

Acknowledgements
Thanks to Barry Attoe, Helen Dafter, Martin Devereux, Jamie Ellul, Clare George, Louise Todd, Zoe van Well and Claire Woodforde of The British Postal Museum & Archive and to the staff of the British Architectural Library, RIBA, for their assistance, suggestions and expertise. Special thanks also to Anna Flood for her valuable assistance in retrieving the BPMA's collection of architectural drawings and to Deborah Turton for commissioning this project, and for her guidance, proof reading and editing.

The British Postal Museum & Archive acknowledges the financial support of Royal Mail Group and thanks Martin Gafsen especially for his personal support, without which this project would not have been realised.

Illustrations are acknowledged as follows:
All images © Royal Mail Group Ltd, 2010, courtesy of The British Postal Museum & Archive, www.postalheritage.org.uk; with the exception of the following which are acknowledged and reproduced with permission:

British Architectural Library Photographs Collection, pages 2, 17 (right), 18, 21, 22 (bottom), 50, 53, 58 (top), 122 (left) and 129; Mark Heathfield, page 25; Tony Hisgett, page 111; Karin McVicar page 116; Steve Mitchell, pages 10 and 124; Nottingham City Council & www.picturethepast.org.uk page 7; Julian Osley, pages iv, 22, 23 (left), 29, 35 (top), 113, 114, 118 (top and bottom), 119 (top and bottom right) and 125; Plymouth Library Services, page 9; Andy Savage, page11; Julian Stray, pages iii, 39 (top) and 70 (bottom); Deborah Turton, pages 34 (top and bottom), 35 (bottom), 43, 62 (top), 108, 110, 112 and 119 (bottom left); Dave Webster page 117; www.fionabird.com page 120.

Contents

- 1 Introduction
- 3 Grand Designs:
  Nineteenth-Century Post Offices
- 47 Modern Post Offices:
  From Post Office Georgian to Standardisation
- 83 Post Office Interiors:
  Towards a Brighter Post Office
- 109 An Architectural Legacy:
  Post Office Buildings Today
- 122 List of Known Works by Principal Post Office Architects
- 132 Publications
- 133 Places to Visit
- 134 Index

The General Post Office. London

Introduction

A post office can be amongst the grandest of buildings in a city. Such buildings are not the familiar 'local sub-post office', typically found amongst neighbourhood shops and run as a franchise, often combined with a newsagent or general stores. These grander buildings, typically located in town centres, are 'crown offices'. Where still used as post offices, these buildings provide a full range of counter services from a dedicated public office, sometimes with a sorting office to the rear. Where no longer used by the Post Office, former crown offices can be found serving new needs as commercial offices, flats or other public entities. Unlike the more familiar local sub-office, crown post offices are or were exclusively devoted to post office business and in most cases were purpose-built to house that service.

To understand the history and development of these buildings it is helpful to understand the background to the Post Office itself. From its seventeenth-century beginnings until it became a public Corporation in 1969 the Post Office was part of the Civil Service, with a government minister, the Postmaster General, at its head; it was essentially controlled by the Treasury. At one stage in its long history, the Post Office controlled all the means of communication (letter and parcel delivery, telegraph and telephone services), as well as offering financial services through its savings bank. Only in 1981 did this monopoly end, with the establishment of British Telecom as an independent company. The postal business continues today under Royal Mail Group, the parent company to Royal Mail, Parcel Force Worldwide and Post Office Ltd.

Crown post offices were for the most part designed by architects who spent all their careers in the Government's Office of Works, and its successors, the Ministry of Works, and the Ministry of Public Building and Works. In many cases because the architects were civil servants little is known about them: often when their buildings were published in the professional press, attribution was given to the Ministry's chief architect. Where the architects were members of their professional association, the Royal Institute of British Architects, more information is forthcoming from their applications to become either Associates or Fellows.

This publication celebrates the work of these previously unacknowledged public sector architects and illustrates their contribution to the UK's architectural heritage.

Opposite: Opened in 1829, the new General Post Office, St Martin's-le-Grand, London, was designed by Sir Robert Smirke, the architect responsible for the British Museum. It provided a public counter and housed mail guards, an armoury, sorting and delivery staff, administration facilities, and was the residence of the Postmaster General. It was demolished in 1912-13.

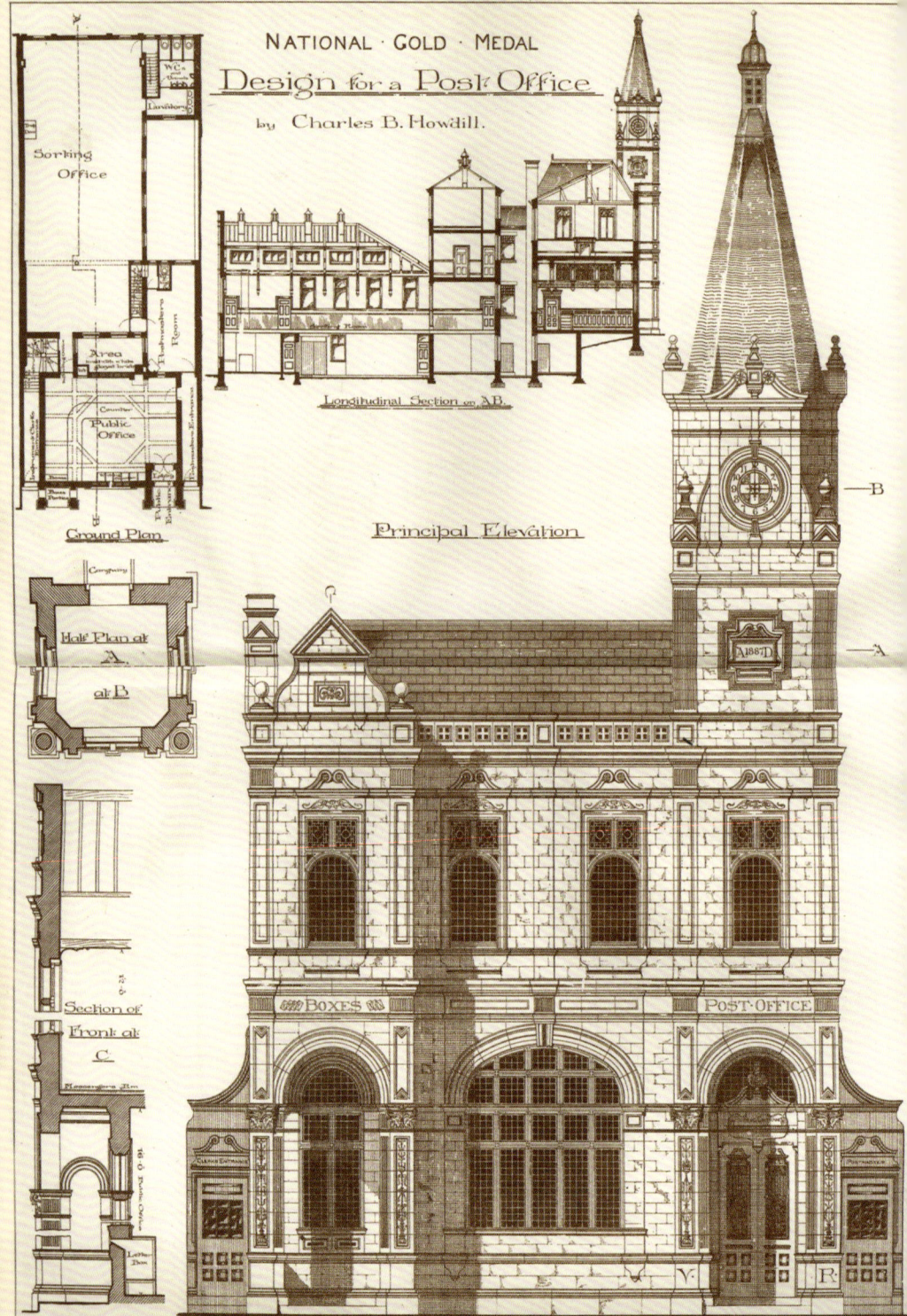

1

Grand Designs

NINETEENTH-CENTURY POST OFFICES

In 1680 the only place in London at which mail could be posted was the office of the Postmaster General in Lombard Street. At this time there were only seventy-seven workers employed by the Post Office in London, and only 316 Post Office staff in the entire country. As the General Post Office (GPO) expanded and became an increasingly important institution, however, larger buildings were needed, not only in London, but across the UK and Ireland.

The first major purpose-built post offices were constructed in the principal centres of postal activity (London, Edinburgh and Dublin) during the early nineteenth century. GPO Dublin was built in the Greek Revival style, fashionable at the time for large public buildings, to the designs of Francis Johnston (1760-1829), architect to the Government's Board of Works. The foundation stone was laid in 1814, and the building opened for business on 6 January 1818. Viewed from the outside the principal feature was the central pedimented portico, supported by six Ionic columns, and topped by three statues, including Mercury, the winged messenger of the gods, a subsequently reoccurring motif in one form or another on many post office buildings.

Opposite: Nineteenth-century post office design was largely the preserve of architects employed within the Government's Office of Works. Private sector architects did occasionally however have an opportunity to prepare a design, as illustrated by this National Gold Medal winning Design for a Post Office, Building News, 1889.

Below: The General Post Office moved to Lombard street, London in 1678 following a temporary relocation forced on it by the Great Fire of London.

The General Post Office, Lombard Street 1780.

In Edinburgh, architects Archibald Elliott (1761-1823) and Joseph Kay (1775-1847) built the post office at Waterloo Place between 1818 and 1819 as part of the development of the North Bridge. The building was a symmetrical Classical design, serving as Edinburgh's main post office until the 1860s, when the service moved to a much larger building in the High Renaissance style, thereafter enlarged and extended.

Joseph Kay had been appointed architect to the Post Office, on the death of his predecessor John Thomas Groves (c.1761-1811), but his designs for a new General Post Office in London were not realised. A new building was required because by the beginning of the nineteenth century the offices at Lombard Street had become excessively cramped and crowded. The St Martin's-le-Grand area was identified as a suitable site because it was more central and because it afforded an opportunity to demolish the slum dwellings, dangerous back alleys, and houses of ill-repute in a notorious area of the capital. Kay prepared preliminary designs, but in 1815 a parliamentary committee proposed that an open competition be held for fresh designs. It was not until 1819 however that the competition was finally held, with Kay's original designs kept secret. Of one hundred designs submitted, none were regarded as suitable. In 1823, a panel of the most important 'official' architects of the day, John Nash (1752-1835), Sir John Soane (1753-1837), and Sir Robert Smirke (1781-1867) finally examined Kay's designs, but found them wanting. Consequently, because the proposed site was in the area for which Smirke was responsible, the Treasury commissioned him to design the new building.

GPO Headquarters moved to St Martin's-le-Grand in 1829. Smirke's design was in the Greek Revival style, with a giant central portico. Inside was a large central hall, from which offices for inland and foreign letters were reached. The building contained 190 rooms on five floors; however additional space was soon required to cope with the enormous expansion of business. During 1837-8 Smirke enlarged the twopenny-post office; in 1845-6 he designed additional rooms for the sorting and letter carriers' offices; finally, in 1892 an extra storey was added. Although Smirke's building was praised and critiqued in equal measure, 'as a piece of purposive design it was superb. Its rational planning and formidable structure stood the strain

Opposite: GPO Dublin, opened in 1818, is one of the oldest and most celebrated of post offices. In April 1916 it became the headquarters for the Easter Rising, the catalyst for Irish independence. Much destroyed, it took thirteen years to restore to its original glory. Today the six fluted Ionic columns, 54 inches in diameter, remain pocked with bullet marks.

Below: Engraving of the New Post Office, Waterloo Place, Edinburgh.

of public business for the best part of a century'.[1] As such, its eventual demolition in 1912-3 caused a public outcry.

Throughout the rest of the UK and Ireland, letters were sent and collected via a growing network of receiving houses in urban and rural areas. These were typically small shops or public houses. Residents of large towns would collect their post from the nearest receiving house, those in less populated areas would either collect in person from further afield or pay a local postmaster or other individual to deliver it to them. By the early nineteenth century, and particularly after the introduction of the Universal Penny Post in 1840, however

Left: Engraving, West Country Mails at the Gloucester Coffee House, Piccadilly. *Prior to the widespread provision of dedicated buildings, mail would be delivered to and collected from a network of receiving houses, including a number of coaching inns and coffee houses used by mail coaches.*

Opposite: Nottingham New General Post-Office, Illustrated London News, *November 1847.*

business grew to such an extent that larger and more sophisticated buildings were required for the efficient running of the postal service.

Before the introduction of the Penny Post, regional post office accommodation was not always of the highest standard. Birmingham's early post office was described as 'a quiet, unpretending private house, with a small one-storey room adjoining... rustic and unbusiness-like'.[2] William Lewins described the typical provincial post office of the 1830s as:

> Situated in the most central part of town, the outside of the building partaking of the ugly and old-fashioned style of the shops of that day. It was then considered quite sufficient for the business of the place that there should be a small room of about twenty feet square devoted to postal purposes; that there should be a long counter, upon which the letters might be stamped and charged, and a small set of letter-boxes for the sorting processes.[3]

Prior to the advent of postage stamps and subsequent services, the public visiting a post office required little attention. Communication was conducted through a trap-door, or aperture, in a wooden pane in the office window, as confirmed by W. Clarke in his memoir of Leamington Spa postal services in the nineteenth century:

> A more primitive, dark and dismal place to dignify by the title of Post Office can scarcely be imagined. There was a mean little porch to it, supported by two pillars, and the entrance was by two of the narrowest of doors. Inside the visitor found himself in a narrow wooden passage without a vestige of light, and, knowing the custom, he would tap on a little panel in the wall, up would fly a little shutter, and out would pop the head of the Postmaster...[4]

By the 1840s, however, post office accommodation in urban areas was improving. The need for larger premises followed the introduction of Rowland Hill's uniform basic postage rate, and the resulting rapid increase in use of the postal service. Larger offices began to provide an interior space for the public, with a counter, instead of the forbidding

aperture. New offices became a focus of local interest and were frequently reported in local newspapers, as with the new post office in Leicester:

> It comprises a large apartment for the receipt and sorting of letters, a room for the Postmaster, a room for the deliverers and messengers, a sitting apartment, offices, and a wide passage for the shelter of the public. In the latter are windows where applicants for stamps and letters and money orders may be waited upon. We believe the officers of the establishment will find their operations very much facilitated in the new building; while the public will reap the advantage of greater convenience and dispatch...[5]

That same year, the foundation stone was laid for a new post office in Nottingham. This was a grand occasion, involving a procession headed by the Mayor in full regalia, accompanied by local dignitaries, and the burying of a hermetically sealed bottle containing specimens of goods manufactured in Nottingham. In his speech, the Mayor referred to the rapid increase in business that had made the building of a new post office necessary, and suggested that the office would contribute to the prosperity of the city's manufacturing industries and commercial interests.[6]

At this time, the Classical or Italianate style was popular as, for example, at Devonport where the first post office opened in 1849.

Opposite: Engraving of the New Post Office, Devonport.

It was designed in a Classical style by local architect George Wightwick. The building boasted a rotunda and four Corinthian columns, with a sculpture of the Caduceus of Mercury, the symbolic representation of the post office, recessed into a panel on the first floor, and bore an inscription said to have read:

> *This Post Office - to embellish the town and to advance the progress of art, is dedicated to the use, and freely entrusted to the care of the inhabitants, in the full belief that all will respect the motives of the founders, and abstain from any act which may disfigure the building, or be repugnant to propriety; so that this tablet may long continue to prove that unreserved confidence in the people will never be abused.*[7]

Premises were often leased by the Post Office from landowners or property developers, resulting, over time, in many towns and cities having head post offices on more than one site. For example, when Sheffield's Angel Street post office proved too small after just four or five years the landowner, the Duke of Norfolk, agreed to build a new post office in Market Street in return for a twenty-one-year lease at £100 per annum. In Doncaster in 1859, post office premises were taken on a thirty-year lease at a rental of £80 with about £550 spent on alterations.

By the 1860s post office premises had considerably improved:

> *...into what a grand establishment the Post-Office itself is metamorphosed! The part now dedicated to the public might be part of a first-class banking establishment. Entering by a spacious doorway, with a lofty vestibule, there is accommodation for a score of people to stand in the ante-room and leisurely transact their business. Then there runs along the whole length of the first or public room a substantial mahogany counter, behind which the clerks stand to answer inquiries and attend to the ordinary daily business.*[8]

A survey of head post office premises in the larger towns of Great Britain and Ireland carried out in 1857-8 reflected how varied and complicated arrangements were, suggesting that the quality of accommodation depended on the efficiency of these arrangements. The most satisfactory public provision was where the building was Crown property with interior fittings owned by the Post Office: examples included Aberdeen, Belfast, Glasgow, Inverness and Liverpool. Many buildings were leased, with fittings owned, by the Post Office as at Bath, Birkenhead, Bristol, Devonport, Hull, Leeds, Manchester and Nottingham. These offices generally received a clean bill of health, with one or two exceptions. The most common practice, however, was for the local postmaster to own the fittings, but hold the building on a lease of perhaps twenty years. This category attracted the most criticism, with the surveyor recommending that a more appropriate arrangement would be for the Post Office to provide both office and fittings.

There was a problem, however, with the Post Office assuming more responsibility for premises because it did not employ in-house architectural expertise, nor were its surveyors qualified to oversee construction work. To resolve this, the Post Office asked the Treasury to sanction the appointment of a travelling post office surveyor to

oversee building work. The Treasury's response was to have far-reaching implications: it determined that all such work take place under superintendence from the Government's Office of Works, with any work executed approved by the Post Office.

Until this point the Office of Works, the government department responsible for the design, construction, and maintenance of major public buildings, had only limited involvement with the design and construction of post office buildings – in Dublin, Edinburgh, and London. The Treasury's decision meant that three government departments would now be involved in the funding, design and construction of post offices. As a result, project completion was liable to considerable delay. From now on, if the Post Office needed a new office in one of the larger towns or cities, the Office of Works had first to be consulted as to plans, the suitability of which would then to be referred back to the Post Office. The Treasury would then be consulted and any objections discussed. When finally, after many years, the building was agreed, it was invariably built by a local contractor whom the Post Office might have appointed in the first instance. A further complication was that any building estimated to cost over £1,000 was subject to Parliamentary approval. Premises built and maintained by the Office of Works under these arrangements were designated Class I Crown Offices.

One of the finest examples of premises built by the Office of Works a few years after it assumed responsibility for Class I offices was Newcastle-upon-Tyne post office in St Nicholas Street (1871-4), a four-storeyed building in the Classical style, its elevation embellished with pilasters and columns. The architect was James Williams who favoured a Classical, Renaissance, or Italianate manner. Williams was also responsible for post offices built in Bristol (1868), Derby (1869), Wolverhampton (1872), Wakefield (1876), Hull (1877), Stockton-on-Tees (1880), Scarborough (1881), Manchester (1884), and the Central Telegraph Office (GPO West) at St Martin's-le-Grand, London.

Not all of Williams's buildings have survived, the principal casualties being the post offices at Spring Gardens, Manchester (demolished in the mid-1960s), and the Central Telegraph Office at St Martin's-le-Grand. GPO West, as the Telegraph Office was

James Williams was responsible for a number of important post offices, including Newcastle-upon-Tyne (opposite), Derby (above) and Stockton-on-Tees (overleaf, left). Few records pertaining to the Office of Works during his time there survive, making it difficult to compile a comprehensive list of his buildings. Northampton post office (1871, overleaf, right), for example, displays all the characteristics of a typical Williams design.

For both Oxford (opposite) and Plymouth (left) post offices, Edward Rivers moved away from the Classical tradition, preferring instead to draw on Gothic and Early English antecedents.

also known, opened in 1874. It was of granite and Portland stone construction, ornamented with Doric and Corinthian pilasters. In reviewing the published design, *The Builder* commented:

> It is scarcely to be denied that the design of New Post Office, as published, is of a common-place type. Unless good architectural skill be brought to bear on details, the general proportions and the profiling of mouldings, we may have a damaging eyesore that will last long years.[9]

Given the importance of the building and site (so close to Smirke's General Post Office building), there was some criticism that the commission had not been awarded to a higher profile architect, rather than a little-known civil servant at the Office of Works. It was intended as a GPO headquarters building but, by the 1880s, the Central Telegraph Office had taken over the space, such was the growing popularity of telegraphic communication. It was on the roof of this building that Marconi demonstrated his system of wireless telegraphy in 1896. Because of the nature of its business the Central Telegraph Office was targeted in both World Wars, with major damage caused in December 1940, when the interior was totally destroyed. The building re-opened in 1943 but, as the telephone became the dominant form of communication, it was eventually closed, and demolished in 1967.

By contrast, Williams's High Renaissance design for Bedford Street post office, Covent Garden, was highly commended although the initial designs may have been prepared by an assistant architect in his office, Edward George Rivers. Rivers's other work includes St Aldate's Oxford (1879), Hereford (1881), Plymouth (1884), and the enlargement (1887-90) of Williams's Bristol post office. A report of Rivers's speech at the foundation-stone laying ceremony for Plymouth post office, gives an insight into how the planning of post offices was approached: practical issues were of paramount importance.

> He was no means inclined to put ornamentation in the first rank. There were other considerations of primary importance [such as] the sanitary arrangements necessary to protect the health of the persons employed in the building.[10]

Left: Edinburgh head post office, photographed in 1935; the head post offices in Scotland's two major cities were both designed in the Scottish Office of Works by Robert Matheson.

Opposite, left: Newly built post offices were not always required. The post office at Norwich was converted from a bank building originally designed by Philip Hardwick.

James Williams's counterpart in the Office of Works in Scotland was Robert Matheson. Like Williams, Matheson designed in a solid High Renaissance or Classical manner, his major works being the head post offices in Edinburgh (1865), and Glasgow (1878). In common with nearly all the head post offices built in the nineteenth century both proved inadequate for the demands placed upon them, and were extended and enlarged by Matheson's successors at the Office of Works in Scotland, Walter Wood Robertson and William Thomas Oldrieve. Matheson's other post offices include those in Dundee, Paisley, and Perth in the 1860s and Aberdeen and Leith in the 1870s; not all have survived.

Whenever the Office of Works was not involved in the provision of new premises (it did not always agree to requests to do so), the Post Office adopted a number of methods to upgrade its facilities, including occupying new buildings designed by local architects in private practice, such as Truro post office, designed by Sylvanus Trevail. Another approach was to take over an existing building, a notable example being the post office at Norwich, converted from a bank built in 1866 to the designs of the architect Philip Hardwick.

Occasionally an architectural competition was held to design a post office. One of the most publicised was for the office at Ipswich, won by the architect John Johnson, a regular entrant to competitions in the nineteenth century. The competition for the office at Knutsford, won by William Owen, became the subject of correspondence in the technical press due to charges of plagiarism of a building by a more fashionable architect, Richard Norman Shaw.

Although considerable progress was made in providing new buildings in the major towns and cities, there was still much to achieve. Citizens were anxious that the image of their towns was not adversely affected by the appearance and condition of the head post office. In 1886 Southampton town council, arguing the case for a new office, described the existing building as 'the meanest, dirtiest and shabbiest structure in a main street'. They desired a new building 'of such architectural character as will mark the post-office one of the important buildings of the town and be suitable to the importance of the borough'.[11] It was another eight years before a new post office was opened.

Below: Occasionally, a post office design was selected from entries in an architectural competition, such as William Owen's proposal for Knutsford post office.

Overleaf: The post office at Southampton was built after complaints that the existing building was not fit for purpose and poorly reflected the image of the city.

A more serious issue was public health. In 1881, *Building News* reported an outbreak of typhoid fever, 'generated by the unsanitary condition and overcrowded state of the Exeter Post-Office'. The Postmaster, fearful no doubt for the future of his livelihood, refused to admit the sanitary authorities to inspect the building, claiming that the post office, as a Crown property, was not subject to the Public Health Acts, even though the state of affairs in the telegraph department was declared to be so bad as to be of danger to the health of the city.[12]

By the end of the nineteenth century it was proving difficult, if not impossible, for the Post Office to provide sufficient facilities for the growing urban populations and the enormous increase in business, generated by the expanding range of services. The Post Office Savings Bank, a wide range of licences, telegram services and postcards with an imprinted postal rate and more were all available by 1870. A major new service requiring amendments to many buildings was the introduction of the parcel post in 1883.

By the last decade of the nineteenth century, the system whereby premises were provided by local postmasters was also increasingly unsustainable: frequently postmasters had to invest all their available capital in supplying premises, forcing them to borrow money to provide the service. As a result, in 1895 the Post Office was authorised by the Treasury to provide fittings and furniture for these offices. Two years later it was arranged that this category of premises

Opposite: Francis Masey's Tite Prize Design for a Post Office, Building News, *March 1888.*

(henceforth known as Crown Class II offices, sometimes as Half Crown offices) should be taken on lease by the Postmaster General instead of by local postmasters. Where taking out a lease on an existing building was not practicable, developers or landlords were found to build offices to designs approved by the Postmaster General.

Consequently, there was now a division of responsibility in the provision of post office buildings with the Office of Works maintaining a monopoly over the design of prestigious Class I offices, while Class II offices were the preserve of the Post Office. The former arrangement caused a measure of resentment among private sector architects. *Building News* remarked that post offices produced by the Office of Works seemed to 'come out of one mould', that 'a little wholesome variety might well be infused into their architecture' and that 'this variety would be obtained by putting a few of the large works out to competition'.[13]

Occasionally aspiring architects did have the opportunity, on paper at least, to try their hand at designing a prestigious post office. In 1888 *Building News* published the results of the Royal Institute of British Architects' Tite Prize competition, the subject of which was to design a post office in a large provincial town.[14] The following year, the same magazine published a design by Charles B. Howdill for the National Prize Medal Design for a Post Office. The brief was to design a post office for a town of 30,000 inhabitants, complete with clock tower to mark the position of the office.[15]

When James Williams retired in 1884, he was succeeded by Sir Henry Tanner, who was to become the most well-known of all the Government's post office architects. In common with many architects in the late-nineteenth century Tanner moved away from the Classical or Italianate designs of his predecessor in favour of a more eclectic approach, although his personal preference was for the Northern or Flemish Renaissance tradition, with rooflines enlivened by gables, chimneys and turrets. The North of England is particularly rich in examples of Tanner's buildings, notably at Forster Square, Bradford (1887), where the purchase agreement specified that the area in front of the building be maintained as an open space, and City Square, Leeds (1896). According to *Building News*, the initial designs

PRINCIPAL FRONT, TITE PRIZE DESIGN FOR A POST-OFFICE.
By FRANCIS MASEY.

Right: Post office buildings often featured on picture postcards during their golden age, such as this view of Henry Tanner's Bradford, Forster Square office.

Below: The most flamboyant of Henry Tanner's designs was the head post office at Liverpool, clearly influenced by the French Château style of the Loire region. This influence can also be seen in his designs for Bradford (right), Birmingham (opposite) and Cardiff (opposite, overleaf).

for Leeds were 'generally disapproved of as inadequate and unworthy by the press, the architectural profession and the town council'. Subsequent alterations were deemed acceptable, although it was 'a cause for regret that the architect has not arranged for a more suitable position for the letter-boxes than the one indicated'.[16]

Other post offices designed by Tanner in the North included Keighley (1880), Grimsby (1882), Liverpool Eastern District (1883), York (1884), Halifax (1885), Preston (1903), and Sunderland (1903). In the Midlands, Tanner's principal work was the French Renaissance-influenced Birmingham head post office (1891), built to replace premises barely twenty years old. Numerous other offices were designed by Tanner including Southampton (1894), Nottingham (1895), Cardiff (1896) and Wolverhampton (1895-7). In London, Tanner's work included two Post Office headquarters buildings, GPO North (1895) and the King Edward Building (1911) at St Martin's-le-Grand, as well as West Kensington (1904), South Kensington (1909) and the first buildings erected at Mount Pleasant.

It could be argued that the disadvantage of Tanner's approach was that he failed to define a type of building that could be readily identified as a head post office. Tanner's buildings could easily be mistaken for town halls, or other types of public or commercial building. Buildings designed by the Office of Works were, however, predominantly on prime sites, giving the service the prominence the Post Office demanded and deserved. Although Tanner's reputation as an architect has subsequently diminished, he has a significant claim to fame. The *Architect* magazine reported that his major work, King Edward Building in London, reflected:

> his penchant for the scientific side of architecture that he should boldly essay the introduction of reinforced concrete as a method of construction in a public building of the first importance, recognizing the advantages that this method offers in such a structure as a General Post Office.[17]

The building was the first in the country to use the Hennebique reinforced concrete system, the advantages of which were a saving in cost, and the enlargement of available floor space because of the thinness of the walls. Although the site has since been redeveloped, the monumental public office and entrance in Newgate Street remain.

In Scotland, major post offices built during this fruitful period were predominantly designed by Walter Wood Robertson. These also reveal the influence of the French or free Flemish Renaissance manner, for example, at Galashiels (1894) and Dundee (1898); although Tudor Gothic elements adorn Paisley post office (1893).

The Tanner era ushered in a golden age of post office buildings,

Above: Henry Tanner was the most well-known of the post office architects in the later Victorian era. His post office in Victoria Square, Birmingham was saved from demolition in 1973 following a campaign by the Victorian Society. Now known as Victoria Square House, it is today office accommodation.

Henry Tanner's magnificent King Edward Building, London (below) and post office building in Cardiff (opposite).

Galashiels Post Office: 1894

Paisley Post Office: 1893

Opposite: Tanner's counterpart in Scotland, Walter Wood Robertson, designed predominantly in a Free Renaissance manner, as at Galashiels (opposite, above), although Paisley (opposite, below) has an air of Tudor Gothic about it.

This page: William Thomas Oldrieve delivered an equally impressive series of post offices north of the border, the most outstanding being the Scottish Baronial Aberdeen (right), where he completed work started under Walter Wood Robertson, his predecessor at the Office of Works. Oban (below) is a very different, yet equally impressive building.

Left and opposite: Of the principal architects employed by the Office of Works, only William Thomas Oldrieve worked in both England and Scotland. His approach was eclectic ranging from exuberant Dutch/Jacobean at Aldershot (opposite), to Edwardian Baroque at Lichfield (left).

Opposite, overleaf: Hull post office was the work of Walter Pott, probably the finest exponent of the Edwardian Baroque style.

the legacy of which is still to be seen in Britain's towns and cities today. Among the architects working under his direction were William Thomas Oldrieve, Walter Pott, John Rutherford and Jasper Wager, all of whom made significant contributions in the first decade of the twentieth century to the UK's largest post office building and expansion programme.

At Aldershot (1902) Oldrieve created on a corner site an elaborate design complete with gables and turrets in a combined Dutch and Jacobean fashion. His other offices include Hyde (1899), Weston-super-Mare (1900), Swansea (1901), Hinckley (1902), Barry (1903), Burnley (1903), Burslem (1903), Crewe (1903), Lincoln (1903), Lichfield (1904), Burton-on-Trent (1905), Merthyr Tydfil (1905) and Bootle (1905), where the design is more restrained than Aldershot and incorporates elements from the Baroque vocabulary. Oldrieve was also responsible for extensions to post offices at York (1902), Barrow-in-Furness (1903), Hereford (1903), Portsmouth (1903), Norwich (1904), Peterborough (1904) and Leicester (1905).

In Scotland, Oldrieve's designs where less elaborate; he experimented with Gothic, Neo-Georgian and Scottish Baronial themes, the most notable being the magnificent post office at Aberdeen (1907), described in *Bon-Accord* as:

> *Strictly Scottish in its architectural character, the structure is possibly the finest granite Postal building in the country, and adds yet another to those enriching features that of recent years have made the Silver City a joy to the artistically inclined citizen and stranger.*[18]

Other notable examples of Oldrieve's work in Scotland are the offices at Musselburgh (1903), Banff (1906), Crieff (1906), Kilmarnock (1907), North Berwick (1907), Haddington (1908), Lerwick (1909) and Oban (1910).

The designs of Oldrieve's colleague, Walter Pott, are firmly in the Free Renaissance or Edwardian Baroque style, as demonstrated by post offices at Warrington (1906), Birkenhead (1907), Blackburn (1907), Rotherham (1907), Dewsbury (1908), Hull (1909), Scarborough (1909),

Post Office, Aldershot

Arthur Sumpster, Stationer, 16 Union St., Aldershot.

Right: Telegraph aerials were frequently observable on the roofs of post offices providing telegraph as well as postal services, as at Dewsbury, 1908.

Blackpool (1910) and Bishop Auckland (1911). Local press reports on the new Blackpool post office suggest that Office of Works designs were not always sensitive to the local environment:

> So far as architecture is an expression of the community's taste the Post Office does not count whether it be good or bad art, it is thrust upon us; the design is a product of the Office of Works, and we have to take it whether we like it or not.[19]

John Rutherford worked principally in the South of England, for example, in Dorchester (1905), Weymouth (1905), Canterbury (1906), Cheltenham (1906), St Helier (1908), Taunton (1911), Torquay (1912) and the London area, although examples of his work may be found further north at Boston (1906), Hanley, Stoke-on-Trent (1906) and Grimsby (1910). In reviewing Grimsby post office, the local paper welcomed the design of the 'palatial' new building as

> of handsome character, with a Renaissant facade, the whole of the massive stone front being in harmony with the [adjoining] bank premises, whilst yet remaining sufficiently distinct to secure individual attention.[20]

Grimsby post office was opened by the Postmaster-General Herbert Samuel on 28 April 1910. The souvenir booklet published to celebrate the event details the constituent parts of a typical head post office of the Edwardian era. On the ground floor, the public office with the Superintendent's room; to the rear, the sorting office with an annexe for foreign parcel mails, the messengers' retiring room, the telegraph delivery room, cycle and store rooms. On the first floor, the Postmaster's room, writing room, telegraphists' retiring room, lady clerks' retiring room, telegraph learners' room, the linemen's room and the engineers' store. Over the annexe, the postmen's retiring room, sorting clerks' retiring room, a bag store. The upper floors were occupied by the telegraphs – a large open space 'fitted with the latest telegraphic apparatus', with a telephone room adjoining and a battery room and stores.[21]

Dorchester (left), St Helier (below left) and Grimsby (below, right) post offices, each designed by John Rutherford.

Many Office of Works' buildings erected in London in the Edwardian era were designed by Jasper Wager, most notably the Northern District post office in Islington (1906), also the site of the Post Office stores department. Other examples of his work include post offices at Barnet (1904), Woodford Green (1904), Chelsea (1905) and Enfield (1906).

A network of single-storey sorting offices in the London postal districts was also built at this time. Often strategically sited near railway stations, these delightfully detailed buildings in the Baroque tradition may well have been designed by Jasper Wager or one of his colleagues. Examples, not all still in use, exist in Clapham (the most idiosyncratic), Dulwich, Finsbury Park, Hanwell, Kentish Town,

Left: Jasper Wager's post offices are built in the Baroque Revival tradition. His principal work was the Northern District Office in Islington, London, the main feature of the façade being a centrepiece of four caryatids supporting a balcony. It is still in use today.

Opposite and below: In the Edwardian era a series of sorting offices were built in London, close to the railway network. Some remain in use today, for example at Lower Edmonton (top), while others have a new use, as at Winchmore Hill (bottom and below), where the building has been converted to office use following a high quality restoration.

Opposite, overleaf: Another example of Jasper Wager's work, Woodford Green post office, London, 1904.

Leytonstone, Lower Edmonton, New Southgate, Tooting, Walthamstow and Winchmore Hill.

While the Office of Works was busy with its grand schemes in the major conurbations, the Post Office was struggling to deal with the Class II buildings for which it was responsible. Help was forthcoming in 1898 when the Treasury authorised the appointment of an 'unestablished' tracing assistant and draughtsman within the Post Office. However the low wages and lack of prospects meant that a fully qualified draughtsman could not be appointed. Instead there was a succession of young and inexperienced employees, who invariably moved on after training before making a worthwhile contribution to the service.

In 1908, procurement methods for new premises were examined by the *Hobhouse Select Committee on Post Office Servants*. It recommended that the Post Office should assume control for all work done on its behalf by the Office of Works. The following year, the Postmaster General proposed that, because post office design was becoming more specialised, a fully qualified architect should be appointed within the Post Office, as well as a draughtsman with a thorough knowledge of building construction, to ensure that buildings were designed, erected and maintained in the most efficient and cost-effective way.

To the dismay of the Post Office, the Treasury rejected the Hobhouse Committee's recommendations, claiming such changes would result in wasted resources, interdepartmental friction and overlapping of jurisdiction. The Office of Works supported this position by asserting it was not aware of any complaints from the Post Office about the level of service hitherto provided, and that the Post Office could not achieve the same level of efficiency and speed of execution were it to assume responsibility for all post office buildings. Furthermore, the Office of Works' extensive experience of office furnishing and arrangement in other buildings under its control enabled it to make recommendations of service to the Post Office, while resisting its extravagant demands, for example in the provision of cleansing and disinfecting compounds, coal buckets, and telephone chairs, where the Office of Works could supply equally satisfactory

Opposite: Henry Tanner's elevation design for Leeds (bottom), City Square, post office and the building today (top).

products more cheaply than those specified by the Post Office.

Following further representations from the Postmaster-General, the Treasury partially relented and sanctioned the employment of an architectural assistant within the Post Office ('architect' was deemed too elevated) and an 'unestablished' trained draughtsman to assist in the provision, fitting up and inspection of Class II offices only. In agreeing to this appointment, the Treasury made it clear that these employees were to have nothing to do with either the Office of Works or Class I buildings, nor did the Post Office have the authority even to provide fittings and furniture in Class I offices.

The need for such an appointment within the Post Office had been highlighted a few years earlier in Coalville, Leicestershire, where a Post Office-leased Class II building had been abandoned within two years of occupation because of serious health problems caused to staff by defective drainage and the close proximity of a slaughter-house. This had resulted in the loss of £1,500 of public money in the hiring of new premises and defrayment of medical expenses for the Postmaster. It was concluded this could have been avoided had the Post Office had in its employ a trained member of staff with technical knowledge of building construction and surveying issues.

Frederick Palmer (1874-1934) with ten years of experience in the Office of Works, was appointed. It was his principal role to advise on any cases of Class II offices (existing or proposed) with regard to planning, sanitation, specifications, and future extensions. Practical issues were still considered more important than aesthetic considerations. By 1911 Palmer was complaining he was able to cope with the workload only by considerable sacrifice of leisure time and leave, and justifiably so. In the first three years of his appointment the number of Class II offices grew from 700 to 1,100. As a consequence, the Treasury agreed to the appointment of a second architectural assistant, W.H. Ludlow (d.1972).

Palmer's major achievement at the Post Office was the development of a cheaper standard for Class II offices, known as the Ingatestone type, as well as standard designs of fittings. He remained in his post until 1920 when he was appointed chief architect to the National Provincial Bank. Here he gained a reputation for designing

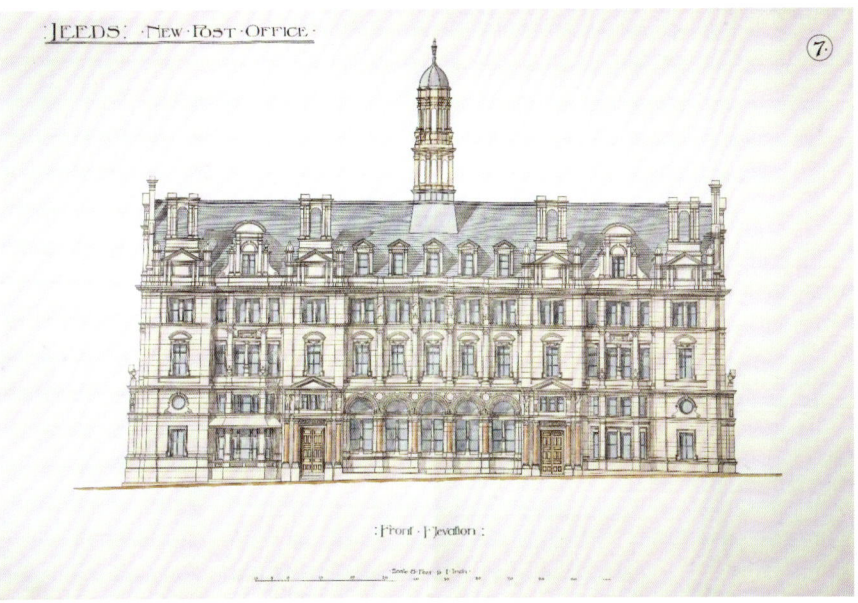

Opposite: Design elevations for a number of post offices including Weymouth (above) and Weston-super-Mare (below) survive in The Royal Mail Archive.

premises sympathetic to local architectural conditions and using wherever possible local materials.

The Treasury's refusal to accept the Hobhouse Committee's recommendations was challenged by the Post Office in 1912 when it took over the services and acquired the premises of the National Telephone Company. The question arose as to which former National Telephone Company buildings be treated as Class I offices, and therefore controlled by the Office of Works, and which be regarded as Class II offices. Following pressure from the Postmaster General, a select committee was appointed to examine the whole issue of responsibility for post office premises.

Sir Frederick Cawley's *Committee on Post Office Buildings* (1913) found no evidence that the Office of Works was unable to carry out the requirements of the Post Office to the complete satisfaction of the Postmaster General. Delays in providing post offices were mostly attributable to the necessity of having to gain parliamentary approval for projects estimated to cost over £1,000. In response to the charge that the Office of Works' standard of construction was too high for Class I offices, the committee acknowledged that the Office, in its desire to attain thoroughness and architectural finish, went to greater expense than was strictly necessary, and recommended that a cheaper standard of building be adopted wherever it was consistent with the size, situation and importance of the office. The Committee was concerned that the Postmaster General would press for an increase in his architectural staff if the Post Office assumed responsibility for all premises, when there was already sufficient expertise available within the Office of Works. After all, post office work represented over forty percent of its staff and budget: removal of this work would constitute a threat to the power and influence of the Office of Works.

The Post Office, however, felt that the magnitude of its operations was comparable to the Admiralty, War Office and Home Office, which were permitted to maintain their own architectural departments. Despite these arguments, the Cawley Committee recommended that the provision of all Post Office buildings be entrusted to the Office of Works, together with the maintenance, supplies and engineering required for those buildings. This would include all the buildings taken

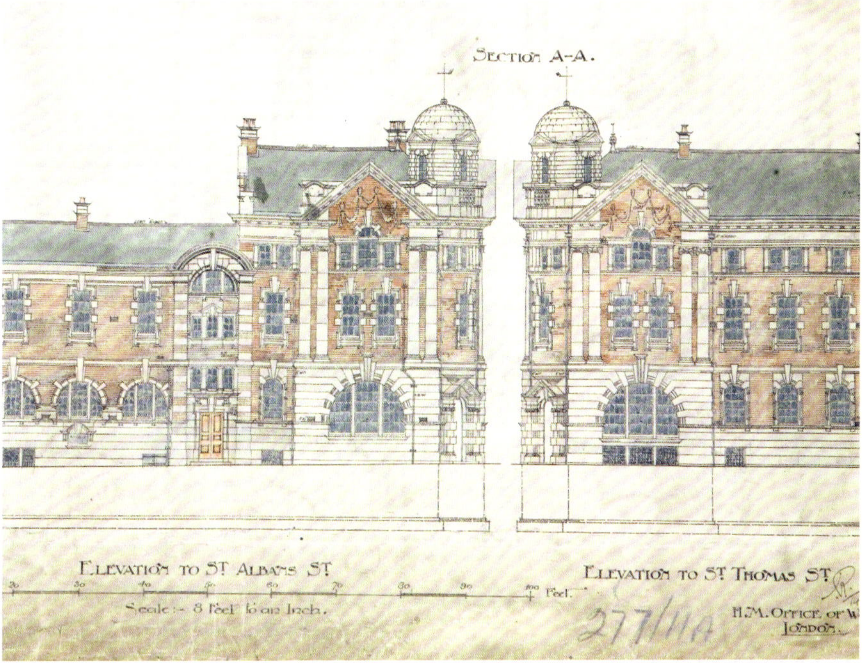
Section A-A.
Elevation to St. Albans St.
Elevation to St. Thomas St.
Scale:- 8 Feet to an Inch.
277/11A
H.M. Office of W
London.

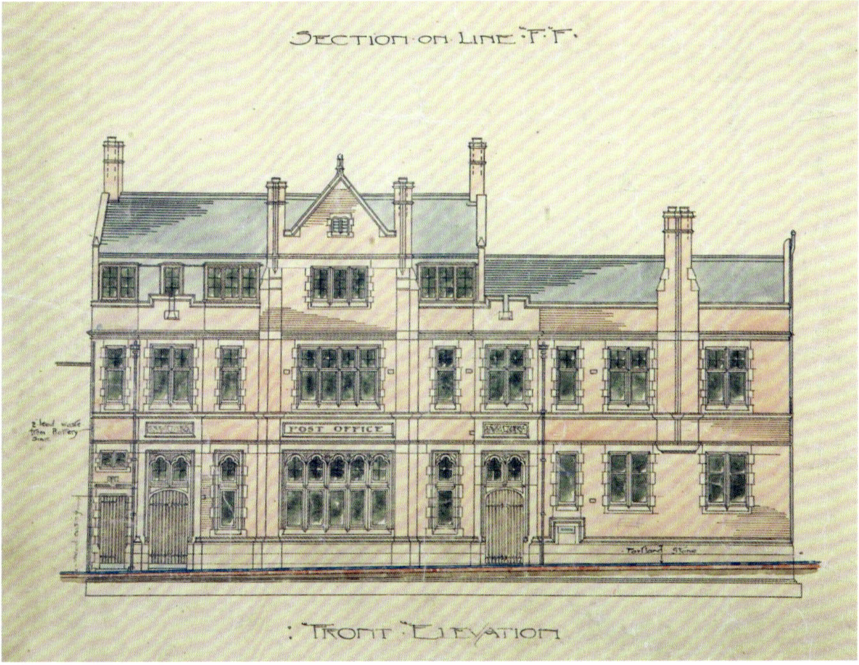
Section on Line "F" F.
POST OFFICE
Front Elevation

Above: Design opportunities were limited for Frederick Palmer, the architect employed directly by the Post Office, although his design for Minehead post office, survives in The Royal Mail Archive.

over from the National Telephone Company, but exclude all offices, categorised as Class III (small sub-post offices). However, the level at which parliamentary approval was needed should be raised to £3,000.

The relationship between the two departments was now established. The Post Office was the client, the Office of Works the architect. The Cawley Committee further recommended that in the matter of design and specification of Class II offices the Office of Works should adopt a cheaper standard than with Class I buildings. The report was issued with a minority submission that supported the Post Office's view that Class II offices should remain the responsibility of the Post Office, and that its architectural department should be strengthened to deal efficiently with construction work.

The Post Office was unhappy with these recommendations. The Postmaster General, Herbert Samuel, declared that he was not satisfied with either the position or the style of post offices in cities and towns which he had visited, with side streets too often being chosen. It was his view that the standard adopted in this respect compared unfavourably with that prevailing in some continental

Left: Ironwork detail outside Jasper Wager's Northern District Office in Islington, London.

Overleaf: At Rochdale, the Corporation made a contribution towards its new post office from its own resources, to achieve a higher quality building.

countries, Germany, in particular, where the position and architecture of the Post Office, in common with other State buildings, assisted in raising German towns above the level of meanness of appearance at which too many of British towns were allowed to remain.

Although the Treasury had declined to accept the recommendations of the Hobhouse Committee, it found it impossible to ignore the Cawley Committee's proposals. In response to the Postmaster General's observations about the quality of Office of Works' buildings, it declared that a serious attempt should be made by the Office of Works to build post offices to a cheaper standard, while at the same time ensuring satisfactory design and proportions for post office buildings. If a local authority desired more costly materials be used or that the buildings be placed on a more prominent site, then Section 49 of The Post Office Act, 1908 made it possible for the local authority to make a contribution. Examples of this practice include Rochdale Corporation's contribution of £2,900 towards the cost of a more elaborate facade for its new head post office and Sheffield Corporation's payment for an outside clock for its new premises.[22] With this new financial reality, the grand designs of nineteenth-century purpose-built post offices were to make way for the less elaborate structures of the inter-war years and later twentieth century.

Notes

[1] Crook, J. Mordaunt and M.H Port. *The History of the King's Works.* Vol. 6: 1782-1851. London: H.M.S.O. 1973.
[2] 'Birmingham Post Office: past and present'. *Blackfriars Magazine*, Oct. 1888.
[3] Lewins, William. *Her Majesty's Mails: an Historical and Descriptive Account of the British Post-Office.* London: Sampson Low. 1864.
[4] Clarke, W. 'The Post Office at Leamington Spa'. *St. Martin's-Le-Grand*, Apr. 1894.
[5] *Leicester Chronicle*, 14 Aug. 1847.
[6] *Illustrated London News*, 6 Nov. 1847.
[7] Available at www.plymouthdata.info/PostOffices.htm. Accessed 16 Aug. 2010.
[8] Lewins. Ibid.
[9] *The Builder*, 24 Dec. 1870.
[10] *Western Morning News*, 17 Nov. 1882.
[11] *Building News*, 19 Mar. 1886.
[12] *Building News*, 23 Oct. 1881.
[13] *Building News*, 24 Feb. 1888.
[14] *Building News*, 23 Mar. 1888.
[15] *Building News*, 5 Apr. 1889.
[16] *Building News*, 24 Jul. 1891.
[17] *Architect*, 7 Mar. 1911.
[18] *Bon-Accord*, 4 Apr. 1907
[19] *Blackpool Times*, 6 Feb. 1909.
[20] *Grimsby News*, 10 Jan. 1908.
[21] *Souvenir to Commerate the Opening of the New Post Office, Grimsby.* 28 Apr. 1910
[22] Chambers, J. *A Few Notes on Sheffield's Postal History*. Unpublished typescript. 1910.

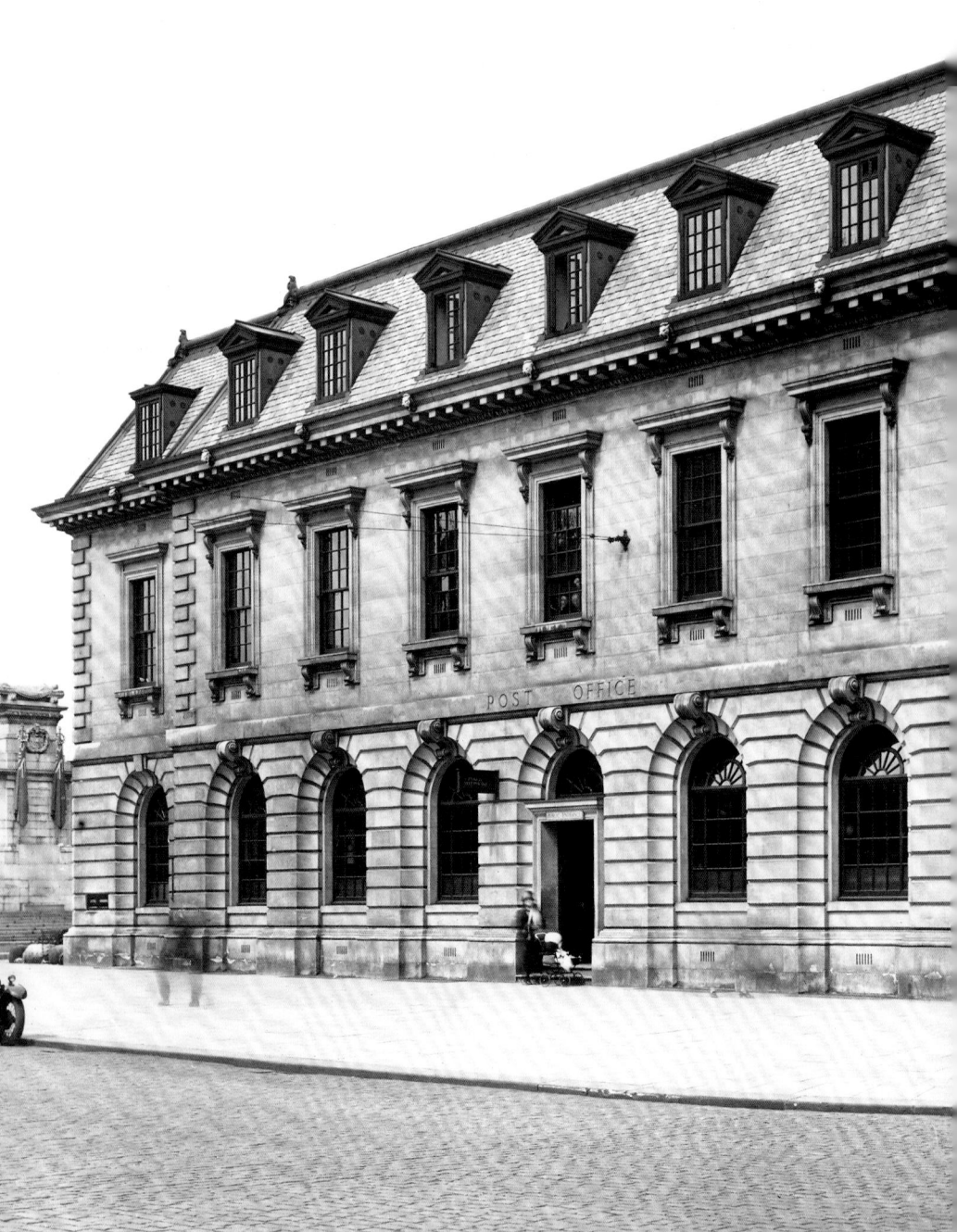

SCUNTHORPE NEW

2

Modern Post Offices

FROM POST OFFICE GEORGIAN TO STANDARDISATION

Post offices built by the Office of Works after 1910 were less ostentatious than their predecessors. By this time, with one or two exceptions, most large towns and cities had been provided with new or expanded post office buildings to cope with ever-increasing demand. It was now the turn of middle-size communities to benefit from the building programme.

Margate's new post office (architect Henry Collins) deployed the now familiar red brick with stone dressings combination in a more Neo-Classical style. At the opening ceremony, as reported in the local paper, Herbert Samuel, the Postmaster General claimed:

> *a post office is a sort of automatic meter which gives a correct measure of the prosperity and development of the place in which it happens to exist... Margate is to be congratulated upon the erection, on a commanding site within its finest public square, of a Post Office admirably equipped in every direction and quite worthy of twentieth century needs.* [1]

Presumably this office satisfied Samuel's claim that the State had a duty to erect public buildings 'in a manner which would contribute to the dignity and the beauty of the towns in which they are situated', particularly 'when those buildings are for the purpose of conducting a vast and very profitable business, and are daily entered by great numbers of people'.[2]

In the smaller and market towns a modest design approach was required. Unlike city post offices, large corner sites or squares were not necessary so long as there was a high-street presence. Fine examples of this new approach were Walter Pott's post office at Malton (1911), built in a red-brick Domestic Revival style, and Charles P. Wilkinson's Scunthorpe (1915) post office. Both foreshadowed what would come to be the ubiquitous post-World War One style: 'Post Office Georgian'.

Wilkinson succeeded Pott as post office architect in the north of England. He enjoyed a prolific career in the Office of Works between 1910 and 1930, creating extensions to major post offices at Stockport (1912), Middlesbrough (1913), Accrington (1920), Southport (1922), Halifax (1923), Blackburn (1924), Preston (1925)

Opposite: Twentieth-century post offices, particularly in provincial towns, were built on a more modest scale than their Victorian and Edwardian counterparts. Scunthorpe post office was an early forerunner of the ubiquitous 'Post Office Georgian' style of the inter-war years.

Opposite: Charles P. Wilkinson's Northwich post office was a lavishly ornamented black and white design in an Elizabethan style, although the theme was not developed in the interior. Such was the novelty of this structure, that it was extensively published in the technical press, the detailing of the central elevation attracting particular interest.

and Manchester Spring Gardens (1926). His designs for new buildings, however, demonstrate unusual versatility.

While Wilkinson's post offices at Huddersfield (1915) and Bolton (1918) adopt a traditional approach in the Classical or Renaissance tradition, his building at Northwich (1914) was unlike any other built by the Office of Works. This remarkable structure illustrates that the Office was no longer happy to disregard the local environment when erecting a new building. Northwich was a special case; there was a history of subsidence caused by the local salt mines. In 1891 the Brine Compensation Act was passed to compensate businesses for subsidence, so long as a special mode of construction was employed to mitigate the effects of mining. As a result, Northwich post office was designed as a medieval timber frame structure, allowing it to be lifted or even moved if required.[3]

The outbreak of war in 1914 meant there was no official resolution to the issue of departmental responsibility with regard to the design and construction of new post offices. After the cessation of hostilities, however, a new spirit of co-operation emerged. There was now a new and more immediate reality: dealing with the difficulties and constraints of the post-war effort. There was also growing public demand for post office services, especially following the introduction of old age pensions (1909), Army and Navy separation allowances (1914), and vehicle excise duties for motor cars (1921).

The decade after the end of the war was characterised by a concerted effort to accept a cheaper standard of construction, initially at the request of the *Reconstruction Committee for Emergency Building*. It was accepted that the most appropriate site for post offices was determined by the location of a town's railway station or commercial centre and that it would be a false economy to locate buildings in more suburban locations because there would be the additional cost in the collection and delivery of letters, and getting mails to and from the station. As for external appearance, enforced austerity demanded simpler elevations, with fewer adornments and embellishments. It was proposed to use brick to a greater extent rather than stone for street elevations and that decorative elements be reduced to a minimum. As a result of post-war timber shortages, some smaller offices were built with flat roofs.

Left: A growing emphasis on minimalism, restraint and economy, is observable in David Dyke's post office and telephone exchange, Leigh-on-sea, built 1925.

Opposite: David Dyke's new post office at Herne Bay, 1935.

The agreed aim now of both the Post Office and the Office of Works was to provide buildings which would blend with the best buildings in the immediate area. For example in a town where early eighteenth-century architecture predominated, a building with a Queen Anne or Georgian elevation would normally be provided. In districts where stone was customary, this material would be used with due regard to economy. Since many provincial towns retained a distinctly Georgian character, by adopting this style a new post office would complement its surroundings and reinforce the image of the post office as one of familiarity, reliability and solidity. With one or two exceptions this policy, directed by Sir Henry Tanner's successor, Sir Richard Allison (1869-1958), was rigorously followed until the outbreak of the Second World War. Although Allison had been assistant architect for the ambitious King Edward Building and had extensive experience working on large public buildings in stone, his personal preference as an architect was the use of brick.

In 1926 Gerald Wellesley described how post offices should be presented to the public:

> *A post-office must be in a prominent position. It should look dignified and permanent, and should, as far as possible, harmonise with its surroundings... The public office, which should, of course, be of a size adequate to the numbers frequenting it, should, in the larger instances, have doors giving on to the street at both ends... The public office must also be very well lit, and this may mean windows on the ground floor which, ideally speaking, are disproportionately large compared with those in the upstairs office. A clock and prominently displayed letter box are also features of a post office front.*[4]

The skills and good taste demonstrated by the Office of Works architects in designing these buildings at a time when resources available for public buildings were severely restricted, were praised by commentators like B. S. Townroe.[5] Equally, the importance of well-proportioned buildings using local materials to produce a

harmonious design was reiterated by the *Brick Builder* which, commenting on country post offices, declared:

> *How important this matter of fitness to environment is will be appreciated by all who have suffered from the constant jarring impressions of incongruous and blatant structures run up with no consideration whatever for the company in which it is to retain. How many restful old market towns and ancient rural communities have been ruthlessly desecrated by a strident, domineering intruder which no ageing or weathering can ever make companionable!*[6]

In 1928 building activity resumed in earnest and it ceased to be necessary to obtain Parliamentary approval for each individual building scheme. A large building programme was begun for the replacement of unsatisfactory, out-of-date, or inadequate premises, and for the enlargement and re-fitting of other offices. The Office of Works continued to build and maintain Class I offices, and in 1938 a self-contained Post Office Section within its Architects' Division was created, while the Post Office retained its architect and surveyor within its Headquarters and Building Supplies Branch to deal with premises for which it retained responsibility.

Of all the Office of Works architects, David Dyke was the most prolific. In his application to become a Fellow of the Royal Institute of British Architects Dyke stated that he was the architect of some hundred buildings, many of which were post offices. He was appointed architect in the Office of Works in 1913 and remained there for the rest of his career. His work is found almost exclusively in the provincial towns of south-east England and East Anglia, and with one or two exceptions, designed in the neo-Georgian manner.

At their most simple, Dyke's designs were rectangular blocks of soft-toned brick with Georgian sash windows, a parapet and hand-made, sand-faced roofing tiles. Dyke's buildings could be sparsely detailed, as at Leigh-on-Sea (1925) or with a minimal concession to decoration with the addition of finials at each corner of the parapet, as at Herne Bay (1935).

Left: Many of David Dyke's post office buildings were in the Georgian style, as at Beccles (near left). Others, including Cambridge (far left) were designed to harmonize with an established urban environment.

Opposite: David Dyke was the most prolific of the inter-war Office of Works architects: most of his work may be found in the South of England. He is most associated with the Office of Works 'Post Office Georgian' style or variations thereof. His work was featured in the architectural press, for example at Uckfield, which was illustrated in Architecture Illustrated, *October 1934.*

The finial motif was common in Dyke's buildings – it appears at Parkstone (1927), Maidstone (1928), Uckfield (1928), Woodbridge (1934), and Chichester (1937) – and was adopted by other Office of Works architects, for example, Archibald Scott at Loughton, Essex (1930), and Archibald Bulloch at Evesham (1926).

Occasionally Dyke would allow himself the luxury of a gable ended roof with chimneys, as at Beccles (1928), or a doorway featuring Classical or Baroque elements, as at Maidstone and Edenbridge (1930), or by introducing stone dressings to window surrounds or a circular window or two, as at Braintree (1929) and Bexhill-on-Sea (1931). At Cosham (1927) Dyke used brickwork of varying tones to introduce an element of decoration. The most interesting of his designs is probably the post office at Uckfield, Sussex (1928), which combines many of his favourite decorative elements.

Dyke's Georgian inspired domestic design worked well for smaller post offices because it was relatively easy to retain a measure of scale and proportion, but was less successful for larger sites such as Maidstone. A passionate advocate of modern architecture, P. Morton Shand, wrote of Dyke's Maidstone office:

> Stumbling upon that exemplar of Office of Works post-bellum elegance, Mr D. N. Dyke's new general post-office at Maidstone, after several years' absence on the Continent, the writer could not help asking himself what had become of the vaulting pioneering genius of our race; and whether the England of today, which this neo-Augustan building was presumably intended to express, is no more than a well-laid-out garden suburb, where the paramount claims of 'atmospherics' are allowed to stifle anything recognizable as a reflection of the spirit of the present age... the threadbare classical symbol of Mercury' winged head is not enough.[7]

ARCHITECTURE ILLUSTRATED.
OCTOBER, 1934.

NEW POST OFFICE AT UCKFIELD, SUSSEX.

D. N. DYKE, F.R.I.B.A., ARCHITECT OF H.M. OFFICE OF WORKS.

Red bricks have been used for the surrounds to the white window frames and also for the arches over the ground floor windows. The walls are faced with brown coloured bricks, stone being used for the main entrance surround and the dressings. The front door is painted green.

Although Dyke worked predominantly in the neo-Georgian manner, he produced successful designs in more urban situations, as at Ware (1928) and later at Cambridge (1935), where the elevation follows the Renaissance tradition. Woodbridge post office presented a different kind of challenge for Dyke, where the building to be demolished to make way for the new post office contained so much interior work worthy of preservation that it was incorporated into the new building.

The impressive quantity of Dyke's output in the 1920 and 1930s demonstrates how committed the Post Office and Office of Works were to providing new post offices (particularly in the south of England). Equally, the quality of his work is matched or even exceeded by his colleagues, Archibald Bulloch, Frederick Llewellyn, H.T. Rees, and Henry Seccombe.

While Dyke's buildings were predominantly brick, Archibald Bulloch experimented more widely – using local stone from the Forest of Dean at Aberdare (1924), and local stone and granite at Bodmin (1925). Bulloch's most challenging and sensitive project using stone was Bath (1926) where, on a wedged shape site, he produced a design in stone incorporating a distinctive Venetian window, the whole sensitively harmonising with the city's eighteenth-century architecture.

Bulloch's colleague in the West of England and Wales, Henry Seccombe, produced equally high quality designs in the 1930s, most notably, at Sidmouth: a red-brick building, with cream limestone dressings, a polished granite plinth, and a slate roof. The most notable features to the street front were the three arched bays supported by Doric columns. This building is now regarded as a 'reminder of the qualities of design and finish possible in public buildings at this period, and a much more scholarly essay in neo-Georgian than is normally achieved later'.[8]

In the interests of harmony, Seccombe's Glastonbury (1938) post office departed from the neo-Georgian theme (which would have been inappropriate) in favour of a Tudor style; while his Maesteg (c.1935) post office recalls a Welsh chapel. Seccombe's other buildings include offices at Welwyn Garden City (1930), Lampeter (1933), Brixham (c.1935), and Malvern (1936), as well as Chipping Norton (1930) and Yeovil (c.1932), where both buildings are faced

Opposite: Although brick was the preferred material, Archibald Bulloch's post office at Bodmin (below) makes use of local stone. His post office at Bath (above) was particularly sympathetic to the city's surrounding eighteenth-century architecture.

Above: The Post Office Georgian style was less suitable for larger offices such as David Dyke's Maidstone post office. The keystone depicts Mercury, messenger to the Gods, continuing the motif first used on eighteenth-century post offices.

Henry Seccombe was probably the most talented of the inter-war post office architects. His post office at Maesteg (opposite, above) was designed to blend with its surroundings. While both Sidmouth (left) and Glastonbury (opposite, below) are listed buildings. New post offices were often a cause for celebration to which local dignitaries including the town mayor were invited. At Glastonbury crowds gathered in the street for the opening of their new post office in August 1938.

Below: Archibald Bulloch's post office at Evesham illustrates how disproportionately large ground floor windows were specified to admit as much light as possible to the public office.

ARCHITECTURE ILLUSTRATED.
JUNE 1941.

ROYAL ACADEMY EXHIBITION, 1941.

PROPOSED POST OFFICE, TRURO.

H. E. SECCOMBE, A.R.I.B.A. (H.M. OFFICE OF WORKS), ARCHITECT.

Opposite, above: Seccombe's designs for Truro post office were exhibited at the Royal Academy in 1941. The display of a post office design at such a prestigious location highlights the extent to which both the design and post office buildings generally were viewed as important contributions to mid-century public architecture.

Opposite, below: H.T. Rees's post office at Skegness.

Left and overleaf: One disadvantage of more harmonious post office designs was that they were not immediately recognisable as public buildings and could easily be mistaken for large private houses, as for example Frederick Llewellyn's post offices at Hatfield, 1936 (left) and Dagenham, 1932 (overleaf).

with Bath stone. He was also responsible for the short-lived reconstruction of Plymouth head post office in 1933, destroyed by enemy action in 1940. Seccombe also had the honour of having his designs for Truro post office exhibited at the Royal Academy in 1941.

It is often difficult to confirm who within the Office of Works was responsible for post offices built in the inter-war years (when published in the professional press they were often attributed to Office's chief architect Sir Richard Allison). There were occasions, however, when the department was happy to reveal architects' identities, as at Herne Bay, where the souvenir brochure published to accompany the opening ceremony included a photograph of the architect, David Dyke. Dyke was also given the opportunity to address the assembled guests, as was Henry Seccombe at Glastonbury when it was formally opened on 17 August 1938.

Two other mid-century Office of Works architects, H.T. Rees and Frederick Llewellyn, also made valuable contributions. Rees worked in the Midlands and North of England, designing post offices including Ilkeston (1920), Gosforth (1923), Chesterfield (1924), Shipley (1924) Louth (1928), Redcar (1928), Skegness (1928), and Northallerton (1929), as well as extensions to Leeds and Middlesbrough head post offices.

Frederick Llewellyn worked in the Office of Works during the 1930s in the London area, designing mainly in the neo-Georgian style, as for example at Hounslow (c.1934) and Greenford (c.1938). At Dagenham (c.1932) and West Drayton (c.1935), however, he took as his inspiration a pre-Georgian domestic tradition. Both of these buildings could be easily mistaken for private dwellings.

Of all the Office of Works architects active during this period Albert Myers faced the most challenges. He worked in the service for over twenty-five years, and like Charles Wilkinson, was active before and after World War One, having to adapt to changing architectural styles over this period. He worked predominantly in the south of England, apart from post offices for Northampton (1914) and Bridlington (1927). His pre-war work included post offices at Rochester (1912), Weybridge (1912), Dover (1913), Tonbridge (1915) and Haywards Heath (1915).

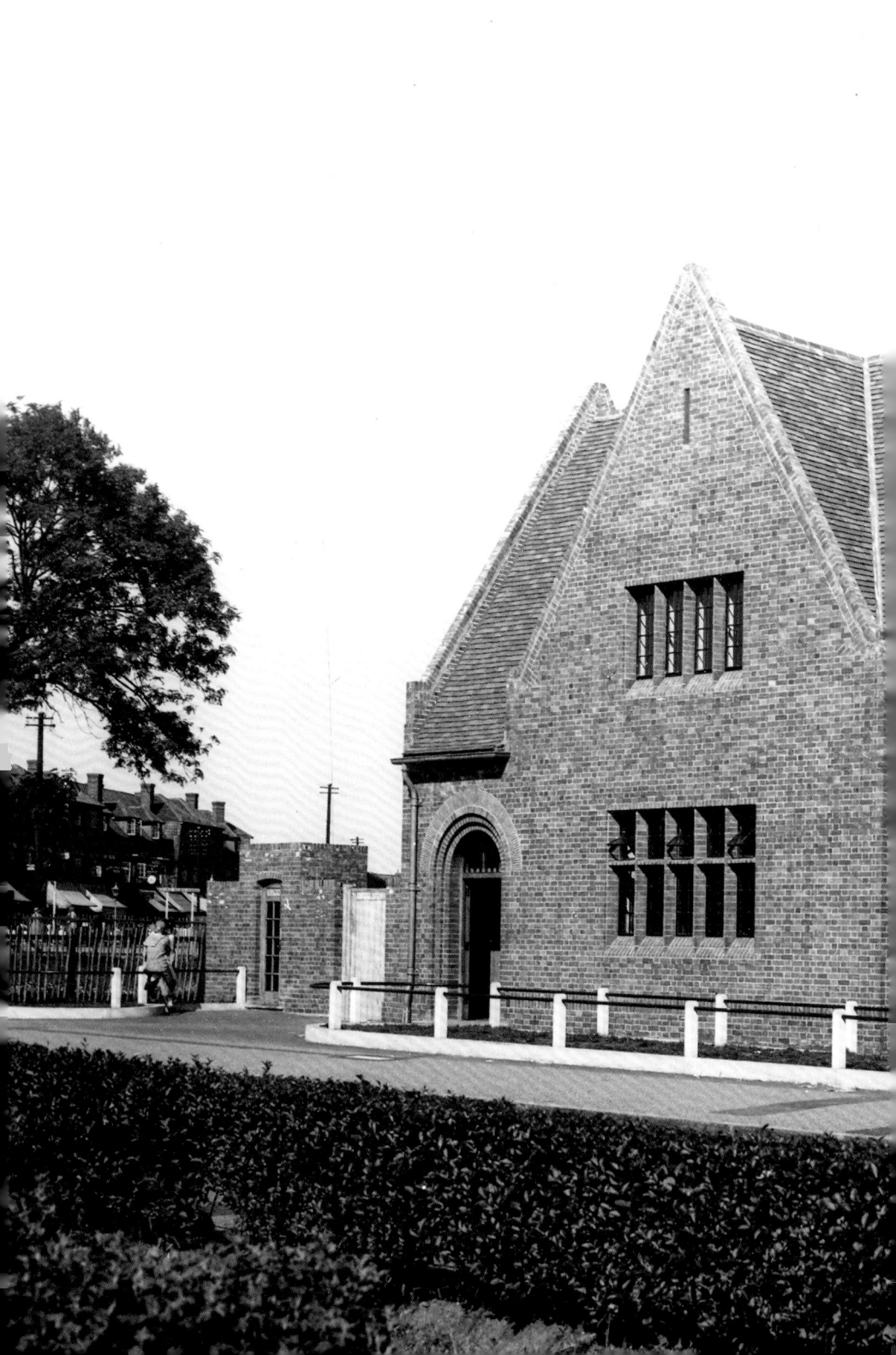

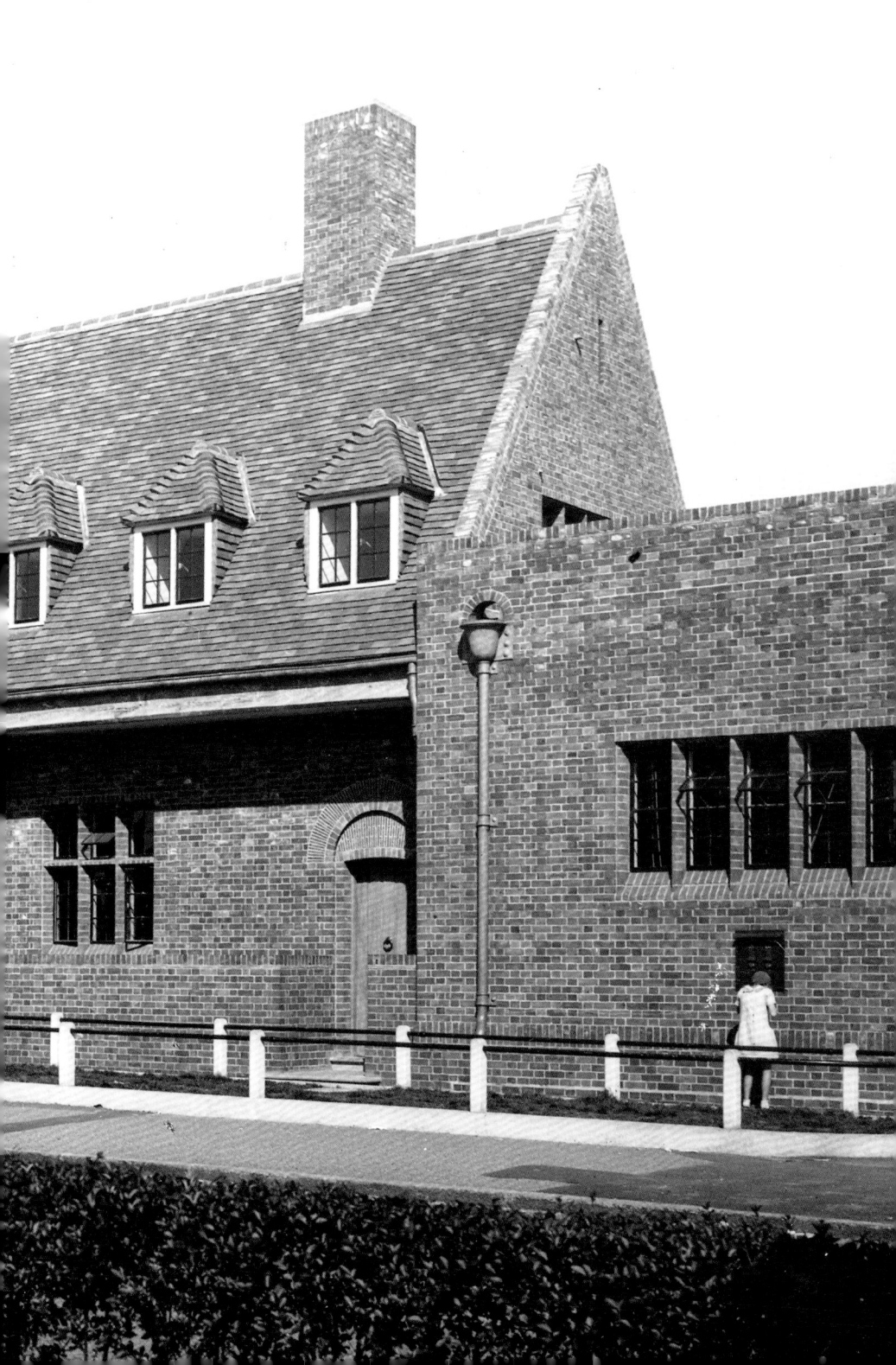

Left: Albert Myer's extensive buildings at Mount Pleasant, London were exercises in modernism that were not repeated to any degree in the rest of the country.

Below: The new post office, Wembley was described in the architectural press as exemplifying 'the modern effort to endow such types of official architecture with a certain vigour in design'. Although attributed to R.J. Allison, chief architect at the Office of Works, the building was designed by Albert Myers.

After the war, Myers demonstrated a more modern approach. His post office at Wembley (1922), was seen as exemplifying 'the modern effort to endow such types of official architecture with a certain vigour in design, which has certainly not always found expression in Post Offices erected during the last twenty-five years'.[9] He also designed one of the first large post-war offices, at Reading (1925), in the neo-Georgian style. In its review of the design, the Reading Society of Architects, while appreciating the quiet restraint and excellence of the design from an academic point of view, regarded it as not forcefully enough suggesting a building belonging to an important department of State.[10]

Myers also designed two major reinforced concrete buildings: Mount Pleasant Sorting Office in North London (1927 and 1937), and Faraday House Telephone Exchange in the City of London (1933).

The latter building was severely criticized for the damage its height inflicted on views of St Paul's Cathedral.

While nearly all new post offices built between 1925 and 1940 were in the neo-Georgian tradition, or variations thereof to suit local conditions, one or two late 1930s designs do illustrate the influence of more contemporary ideas. To address the problem of how to bring more light into the public office, the Office of Works architects devised an elevation which dispensed with individual windows on the ground floor (where the public office was always located), replacing them with a single glazed opening, with steel sashes and stone surrounds. This design was employed at Muswell Hill, London, with traditional sash windows retained on the first floor.

For years continental architects had been experimenting with new shapes and materials. With one or two exceptions, their contemporaries in the United Kingdom had been slow to follow suit, preferring more traditional and conservative designs. In this context, Myers's post office in Penarth (1936) is all the more surprising. It is out of character with the Office of Works model, displaying, superficially at least, features derived from continental modernism: a curved building on a corner site, with a horizontal emphasis throughout and flat roof. An extract from a semi-official Office of Works letter dated 16 December 1936 gives an idea of how the building was finished:

> *The general walling is rendered in cream colour and has a scraped finish. The stone to coping is 'Fishpounds Blue pennant', and to window heads and cills reconstructed blue pennant which is of a blue grey colour. The plinth is a 'green and black' reconstructed 'Impervious' polished granite. Entrance doors are in polished Nigerian Mahogany. Sashes painted a pale blue and the posting box, notice frames and lamps in 'Bronze'.*[11]

In 1939, two other modern-style post offices appeared – at Church Lane, Scunthorpe, of brick construction, with a flat roof and elegant curved frontage, and another at Beckenham, London. Despite the latter's departure from the now familiar 'Post-Office Georgian' in this suburban location, the local paper approved of the building 'erected

Above: Albert Myers designed the first major post office (Reading) to be built after the First World War.

Overleaf: Albert Myers was also responsible for the continental-modernism influenced post office at Penarth, South Wales.

Architect – Mr. A. R. Myers.

in multi-toned bricks, with a sparing use of reconstructed stone... designed in a restrained rendering of the modern style', seeing it as evidence 'of the desire to provide thoroughness and efficiency in every department that the general public will find in the public office'.[12] Beckenham post office looked even more impressive because it was set back from the main road, with wide grass forecourts surrounded by low stone curbing and paved approaches. It was built with an A.R.P. refuge in the basement but after only eleven months of operation was struck by a bomb which demolished the front of the building.

The inter-war post office building programme was perhaps best summed up by Clough Williams-Ellis:

> *The Post Office [sets an] example of considerate good manners, going to real trouble to adapt their several premises to the streetscapes in which they stand... often indeed being the very best buildings, old or new, to be found in many a market town.*[13]

After the war the Post Office took stock of its building requirements. It had six years' arrears in its building programme to make up, plus building losses from the war. Tanner's Liverpool post office had been severely damaged, and new head offices were required to replace those destroyed in Plymouth and Exeter. The Post Office also remained committed to constructing new buildings to meet growing post-war needs in cities including Birmingham and Manchester, as well as smaller towns.

In the early 1950s, the position of architect within the Post Office was transferred to the Ministry of Works (the re-named Office of

Opposite: At Beckenham, like Penarth, the design was clearly influenced by continental modernism.

Left: The large window frontage introducing light to the public office at Cranleigh was designed to render the public office light and bright, while affording maximum visibility from the outside.

Overleaf: Exeter post office, was one of two large West Country post offices designed by Cyril Pinfold in the 1950s, the other was Plymouth.

Works). This meant the design and construction of purpose-built post offices became the sole responsibility of the Ministry.

The immediate post-war period revealed no further dramatic changes in design: indeed, the Ministry's inherent conservatism persisted. Of the few new offices built, Alan Dumble's Woodford Green (1948) post office was of familiar brick construction, rendered more utilitarian by its flat roof and minimal external features. Newmarket post office (1951), replaced the building destroyed by enemy action in 1941, yet looked like a building designed twenty years earlier – complete with Georgian fan-light above the door. It was designed to complement its surroundings, particularly the adjacent Jockey Club building, itself built in 1933, inspired by a Georgian predecessor.

From the 1950s onwards new challenges faced the Ministry of Works architects. Given impetus by the 1951 Festival of Britain, 'modern architecture' increasingly became the dominant style for public sector buildings. A major element of the post-war reconstruction programme was the development of a number of new towns planned to relieve urban congestion in a controlled manner. New towns and those in need of reconstruction after the ravages of war required buildings that looked to the future and which were airy, colourful and light, yet finished in materials that were economical and easy to maintain.

Cyril Pinfold, senior architect at the Ministry of Works, built new post offices at Plymouth (1957) and Exeter (1959), the former a fine example of this new approach. Built on a curved site, and set back to allow adequate daylight to the public office, Plymouth was given elevations to the main street of Portland stone, relieved by panels of slate and touches of colour in the roof canopy, balconies and lettering.

The post office at Cranleigh, Surrey (1960), was fronted with glass, frosted to dado height, and clear above. At Keynsham (1960), the local paper extolled the virtues of the newly opened post office designed by W.H. Ralph, describing it as the 'last-word in Post-Offices'.[14] It was entered by double-glass doors and extremely well-lit, both from the large plate glass windows and artificially.

Thomas Winterburn's five-storey head post office at Luton (1957), reflected the work of modernist architect, Le Corbusier, whose masterpiece of urban living, l'Unité d'Habitation in Marseilles,

Right and below: The post office at Lancing, designed and built after the Second World War, was one of the last manifestations of the unassuming 'Post-Office Georgian' style.

Below, bottom: Although Thomas Winterburn's designs were influenced by post-war modernism, he was also responsible for Diss post office, 1953. Still in use today, its symmetrical design harmonises with its environment and recalls the approach of the pre-war architects.

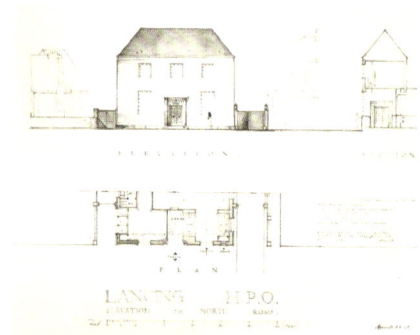

Left: Elevations for later twentieth-century post offices illustrate a growing emphasis on well-lit public offices with ceiling to floor plate glass windows, as with this 1955 design for Favesham head post office.

Overleaf: Thomas, Winterburn's large post office building at Luton was clearly influenced by modern architects like Le Corbusier.

influenced many British architects in the 1950s. Again the issue of admitting light into the public office was addressed by full height windows. This was not the only reason for such transparency – in the 1950s parents were not encouraged to bring their babies into post offices: large windows allowed perambulators to be left outside under full observation. On an equally large scale was Winterburn's sorting office at Norwich (1956), praised as 'a good example of current architecture, nicely grouped, with much curtain-walling... [deserving of] mention in a town so far poor in worthwhile contemporary buildings'.[15]

In terms of construction methods, the traditional 'wet' method of bricks and mortar was replaced by reinforced concrete or steel frames, the structure of which could be hidden by cladding of various materials or by a non-load-bearing curtain wall. The increasing importance of the influence of technology on building was reflected in the development of prefabrication. This was initially employed after the war to speed up the building of houses and schools, later becoming a key element in the production of new buildings, particularly in the public sector.

In 1957 the Post Office and Ministry of Works set up a Research & Development Group to study the possibility of reducing the cost of building telecommunications and post office buildings, by adopting a modular approach to design. Hitherto, buildings had been individually designed; now a standardised approach would be more economical. Another advantage of this approach was that flexibility could be incorporated, allowing for future building expansion or changes in working patterns.

The first post office building to use the modular approach was the telephone exchange at Altrincham (1960), followed by the head post office at Hitchin (1962). The latter opened to a blaze of publicity in February 1962, having been built at two thirds the cost of similar buildings – £64,000 instead of an estimated £100,000. Economies were achieved by critically examining requirements at the planning stage and making some rooms combined-use, thereby making greater use of floor space and reducing the overall area and cube of the building. In addition, the ground floor plan was designed as a shell with demountable internal partitions which could be moved to facilitate future

Left: An important feature of John Parr's post office at Pudsey, Yorkshire, completed in 1957, was the generous sized steel casement windows designed to allow maximum daylight to the public office.

Opposite: Smaller scale Winterburn designs include post offices at Berkhamstead, Canvey Island, Harlow New Town, Saxmundham (pictured), and Stockton-on-Tees.

changes in working practices.

The early 1960s were an interesting time in the post office building programme. 1960 saw the first and only 'drive-in' sub-post office at the telephone exchange, Leicester. Here interesting design challenges presented themselves: the drive-in counter had to be high enough to avoid staff stooping, but low enough for the motorist seated in a car. In November of the same year a flagship twenty-four hour post office at Trafalgar Square (architect Philip Watkinson) opened. This boasted the longest counter in Europe (over 180ft) while a panoramic view of the Battle of Trafalgar provided the backdrop to the parcel posting section.

In 1962 the Ministry of Works was renamed the Ministry of Public Building and Works, taking on responsibility for monitoring the building industry. It continued to provide architectural services to the Post Office. Typical of buildings erected in this decade were E.C. O'Farrell's Kingswood post office (1965). The press release accompanying the opening encapsulated the elements of an ideal post office: a modern clean-cut building conveying an impression of air and light, a spacious public office, and designed to accommodate an anticipated growth of business for the next thirty years.[16]

While architects at the Ministry worked on the Altrincham and Hitchin models, the Post Office also took steps towards modernisation by appointing private sector consultant architects and designers to provide fresh design ideas. Architect Hugh Casson, industrial designer Misha Black of the Design Research Unit, Liverpool College of Art Principal W.L. Stevenson, and Liverpool architect F. J. M. Ormrod, were invited to present ideas for modernising the public office. The brief was to re-design three post offices in London and three in the North of England, applying a different treatment to each, and thereafter over the next eight years to improve 100-150 offices at an annual cost of £300,000. The six pilot offices selected were South Molton Street, Knightsbridge, and Ludgate Circus in London (Casson & Black), and branch offices at Royal Exchange, Manchester, Corn Exchange, Liverpool, and Chapeltown Road, Leeds in the North (Ormrod & Stevenson).

The offices at South Molton Street and Knightsbridge, although

not identical, revealed enough similarities for an identifiable 'house-style' to emerge. The key concept was standardisation, rigorously applied to the smallest detail, thereby achieving clarity of planning, simplicity of form, neatness of detail, and cleanness of colour. Elements of the exterior included a modular metal frame, incorporating a red transom bar, a suspended clock, and centre panel incorporating vending machines, posting box, and information notices.

As a result of these developments, the Postmaster General, Reginald Bevins, announced, at the opening of an exhibition of post office design at the Fleet Building, London, on 19 November 1962, a total change in the design and construction of the 'ordinary' post office: there would be a new 'house-style' based on modern construction techniques. Eric Bedford, chief architect of the Ministry of Public Building and Works, explained in an article in *The Builder*:

> *The basic functions of all post offices were similar, but [in the past] in each new building an architect worked out his building design and decorative scheme from first principles. Valuable professional time was wasted on individual solutions to common problems. The results were often pleasing but did not reflect any common style. The principle behind the new system was that the basic functional*

Left: Prefabricated frontage to Harrow post office, a typical feature of mid-century public sector architecture.

Opposite, above: Kingswood post office, 1965, built at a cost of just over £55,000, was part of what the Post Office called 'a scheme to provide Post Offices which give pleasant surroundings in which the public can do their business'.

Opposite, below: Hitchin post office, 1962, was built to a modular design that eschewed extraneous detail but offered capacity for expansion, while its construction costs were considerably lower than those built on more traditional lines.

needs became the starting point for standard designs. These designs were translated into a system of standard factory-made components which could be assembled in various ways to suit individual offices... This would make for instant recognisability and would provide a 'house-style' which would be associated with the Post Office in the public mind.[17]

In the future, post offices would be planned to a standard module adapted to functional needs and the materials used. Office façades would consist of a number of repeating standard units. The standard panel would be an aluminium frame with service units (posting box, stamp vending machine) in stainless steel. The dominant external feature would be a prominent red illuminated transom bar across the whole façade bearing the words 'POST OFFICE AND SAVINGS BANK'. An additional sign would bear 'POST OFFICE' only. This was based on Casson and Black's work at Knightsbridge post office. The intention was to start constructing new offices in this way by 1965.

The implementation of this ambition, enthusiastically promoted by the Ministry of Public Building and Works (its building programme policy was to encourage industrialised building) was fraught with difficulties from the outset, particularly with regard to cost. It was assumed that mass-produced units would be cheaper, but by 1964 questions were being raised within the Post Office about the economic advantages of adopting the new 'house-style'. It was not certain that savings could be made over more traditional methods of construction; indeed it was claimed costs could increase by eight to sixteen percent. Despite these misgivings, the Post Office Board believed the adoption of a 'house-style' outweighed cost consideration and endorsed the new approach in 1965. Delays in placing contracts for the 'house-style' units, however, meant it was not until March 1970 that sufficient units were available for programme implementation to begin (eight years after the initial notice of intention).

The adoption of the metric system in the building industry in 1970 added further complications because the prefabricated panels had been manufactured in imperial dimensions. They also proved impossible to adapt to fit shop units in shopping precincts where post offices were

TELEPHONES
POST OFFICE & SAVINGS BANK

POST OFFICE & SAVINGS BANK
TELEPHONE
POST OFFICE
Savings Bank
private
no admittance

Opposite: Knightsbridge post office before and after modernisation: whilst the Hitchin model was being developed, the Post Office invited architect Hugh Casson and industrial designer Misha Black to devise and implement inspirational concepts for its buildings and interiors. The result was a near 'house style' incorporating standard units as at Knightsbridge (left) and illuminated red transom bars as at South Molton Street (below).

Opposite: In 1969 the Post Office became a corporation and over the years its traditional ties with the Ministry of Public Building and Works, and subsequently the Property Services Agency gradually loosened. As a result, post office commissions went to architects in the private sector, as at Guildford, 1970-72. Designed by Roman Halter and Associates, the building represented a radical departure, incorporating wrap around glazing and a projecting gazebo.

now being sited. Furthermore, Post Office regional directors were said not to be enamoured of the 'Woolworth-type' image, and local planning authorities did not find them acceptable in sensitive situations. Between 1970 and 1972 standard units were fitted in fewer than twenty offices in the whole country and in 1979 the 'house-style' concept was abandoned.

In the meantime, the Ministry of Public Building and Works had been incorporated into the Department of the Environment, and in 1972 the Property Services Agency was set up to deal with the commissioning and delivery of public buildings, including post offices. By the mid-1970s, the Post Office was concerned about the rising costs of using the new Agency, and the use of private consultants became more widespread, with no further reference to a predetermined 'house-style'. After over one hundred years the domination of the public sector in the design of post office buildings was over.

From the 1980s onwards, priorities and policies changed, and the number of post offices decreased. The Post Office reduced the number of post offices for which it was responsible, while exploring and putting into effect other means of providing its services to the public, for example, by placing offices in high-street retailers including W.H. Smith. Nevertheless, the legacy of the United Kingdom's rich post office building programme lives on.

Notes

[1] *Thanet Times*, 23 July 1910.
[2] The British Postal Museum & Archive. Post 30/3062. file V1. (1913)
[3] *Architect & Contract Reporter*, 26 July 1918; 14 Feb. 1919; 11 Apr. 1919.
[4] Wellesley, Gerald. 'Recent Post Office Architecture'. *Architects' Journal*, 6 Jan. 1926.
[5] Townroe, B.S. 'Post Office Architecture'. *Building*, Mar. 1927.
[6] 'Country Post Offices'. *Brick Builder*, Oct. 1927.
[7] Shand, P. Morton. 'The Post-War Post Office: Recent Tendencies in Post Office and Telephone Exchange Design 1: Great Britain'. *Architectural Review*, Oct. 1930.
[8] English Heritage Listed Buildings Register, no. 403302.
[9] *Architect & Building News*, 8 June 1928.
[10] *The Builder*, 29 Apr. 1921.
[11] The British Postal Museum & Archive. Penarth post office. Portfolio file.
[12] *Beckenham Journal*, 9 Dec. 1939.
[13] Williams-Ellis, Clough. 'County and Country Towns'. *Country Life*, 6 July 1945.
[14] *Bath Chronicle & Herald*, 29 Oct. 1960.
[15] Pevsner, Nikolaus. *North-East Norfolk and Norwich*. Harmondsworth: Penguin, 1962.
[16] General Post Office. Press and Broadcast notice: New post office opened in Kingswood, 1 Dec. 1965.
[17] Bedford, Eric. 'A House-Style for Offices: New Trends in Design and Construction'. *The Builder*, 23 Nov. 1962.

New General Post-Office, London.

INTERIOR.

3

Post Office Interiors

TOWARDS A BRIGHTER POST OFFICE

The evolution of the purpose-built post office from nineteenth-century grand designs to the simpler, more efficiency-focused buildings of the twentieth century was mirrored by similar changes to post office interiors. Having established the need for a public office in which to transact business, the ongoing concern became how best to design an interior which facilitated the efficient provision of postal services to the public.

Familiar features of the public office interior of larger post offices during the later nineteenth century and the first decade of the twentieth century might have included a spacious vestibule leading to the public office; a floor of decorative mosaic; walls sometimes lined with marble; a service counter of polished mahogany in a variety of configurations, often running along the whole length of the counter; a wirework (or later) bronze screen through which business was transacted; a panelled ceiling; and a mahogany telegram writing table divided into a number of compartments. A special supplement of the *Western Figaro* described such features as part of Edward Rivers's Plymouth post office of 1884:

> *The counters and desks are of polished mahogany or wainscot, and there are numerous little places where one can go to write out a telegram or order form, or indite a love letter if need be – there is plenty of room for the stretchers usually told in such compositions. The windows are of plate glass, except in the fanlights, which are of the now-popular cathedral tint; and though simple, the colouring, pale green with white ceiling, is very pleasant. Some ironwork ornament ingeniously conceals the hot-water pipe apparatus, and with a marble top may serve as a refreshment counter, or as a table to take a light luncheon at...*[1]

In celebrating the opening of Robertson and Oldrieve's Aberdeen post office in 1907, the local paper described the public office beyond its carved oak doorway as:

> *... lined with marble to a height of nine feet, and there are two marble-lined pillars in the centre of the office. The floor is of*

Opposite: Henry Tanner's King Edward Building (built 1905-10) in St Martin's-le-Grand, London was until 1998 Britain's principal post office. The public area featured Irish green and Italian white marble, with bronze fittings and a mahogany counter running almost the entire length of the main hall.

ABERDEEN POST OFFICE: 1907

mosaic, the wood-work of wainscot, and the ceiling panelled. The massive mahogany-topped counter forms three sides of a square... Off the public office to the right of the entrance is the telephone call box... furnished with comfortable boxes where the public can speak to other towns. The counter is screened all round with the exception of the part for receiving parcels...[2]

Equally impressive was the interior of Walter Pott's 1910 post office at Blackpool:

Here is a horse-shoe counter, about 70 ft long, for the sale of stamps, postal orders, and general business. Ample table space, for the writing of telegrams and post-cards, is provided, for all along the wall, under the windows, there are no less than 24 compartments for this purpose, whilst round the 2 pillars in the centre of the spacious office, are further compartments for writing purposes. The furnishings of the public office are very pleasing, all the woodwork has been done in polished teak, with the exception of the tops to the counter and desks and these are of polished mahogany. The general beauty of the interior is considerably enhanced by the fine screen on the counter, by which the public is separated from the staff. A well designed 'gauze' of bronze takes the place of the old, familiar wire network, and at the head of each electric standard is an oval section set off with the letters 'E.R.VII', from which rise the globe and electric lights. The walls of this grand

Opposite: Aberdeen post office interior, 1907.

Overleaf: Customers transact business in the public office at King Edward Building, London, early twentieth century.

'sale publique' are lined to a height of 8 feet with faience, and the floor is laid with an excellent design in marble mosaic.³

Henry Tanner's King Edward Building in St Martin's-le-Grand was Britain's principal post office. The interior was palatial. It was the largest in the country, measuring 152 x 52ft, with a counter running the length of the office. The inside walls were lined throughout with marble, a green Irish marble being used for the dado, pilaster panels, door architraves, and the front of the counter, and a light Italian marble, called Arni Alto, for the remainder. The pilasters and piers had bases and capitals of bronze, with bronze also used for the counter edges, and the electric light fittings.

Not all post offices were equipped to the same degree. There was much discussion on how best to fit out public offices. For example, issues including type of counter screen and whether to provide seats and writing tables were regularly debated over the years.

Counter screens were not common until 1887 when the Office of Works and the Post Office agreed that the case for installing a wirework security screen should be judged on local circumstances. The Post Office was, however, in favour of installing screens at all Class I offices. In 1892 the Office of Works suggested that the matter be referred to the Treasury. The Post Office, no doubt fearing that there would be a negative response, was unhappy at this proposal, and continued to press for the installation of screens at least in less salubrious locations, where robberies had taken place, or where women employees worked behind the counter. Although a glass screen was installed at the money order office in Bolton post office, the Office of Works' preferred solution was to make counters high and wide enough to deter theft.

By the time the *Committee on Post Office Fittings* reported in 1911, open counters were still being fitted, although it was common practice to fit a screen of some sort. The Committee felt it desirable that a definite policy be adopted, and recommended that all counters should be provided with screens, except at the point where parcel business was transacted. The Committee also concluded that the practice of carrying out public business at a number of separate counters (or even

Right: Watercolour design for a post office telegram table.

Opposite: Liverpool head post office interior, 1934.

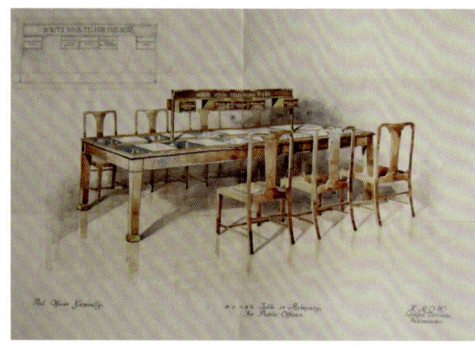

in separate buildings) was wasteful and inconvenient, and recommended that all types of business should be transacted at one counter. While not providing definitive guidelines on how the counter should be configured, the Committee agreed that a 'horse-shoe' design was usually the best, while in smaller offices the counter should run at right angles to the windows. The effect of this arrangement was that the building could be accommodated in narrower high street frontages, making it more economical in terms of renting or purchasing premises.

The Committee's report also considered whether seats should be provided for the public. Hitherto it had been common practice to provide fixed standing units only for those wishing to write telegrams. The report confirmed that this policy should be continued with no further accommodation provided for letter writing, to deter loitering by the public. However, perhaps in recognition of the Post Office's role in local communities, it conceded that in certain areas, such as seaside and other tourist resorts, the public could not be prevented from using the office for the purpose of casual correspondence (this was, after all, the golden age of the picture postcard). The solution was to increase the number of telegram writing desks. On no account were separate chairs and tables to be provided. There was also the issue of whether seating accommodation should be provided for telephone users waiting for trunk calls. Sometimes there was a long wait and complaints were made about the absence of seats. It was accepted that these complaints were justified, and agreed that where space and local circumstance warranted, chairs for trunk callers should be provided if recommended by the local Post Office surveyor.

A subsequent *Committee on Accommodation and Fittings* reported in 1924. By this time the Post Office's preference for counter screens had prevailed. The recommended counter depth was fixed, although the Committee considered that overall the standard dimensions of public offices could be greatly reduced. Despite the length of the report and the characteristically detailed recommendations, little attention was devoted to the public side of the office. Concern was instead expressed about the high cost of enforced enlargements, and the Committee advised that in planning public offices, the Office of Works

should consider the requirements of the next twenty years following an office opening: it would be incumbent on the Post Office to provide the Office of Works with the necessary information as to future developments. Had this advice been taken in the mid-nineteenth century, the life of earlier post-offices could have been extended beyond the average occupancy of fifteen years or so.

From the public's perspective, the typical 1920s post office interior lacked appeal. It was functional, painted in dark colours (perhaps of battleship grey or chocolate brown), with glazed tiled walls, low lighting levels, and the ubiquitous intimidating bronze mesh counter screen. Gerald Wellesley posed the question:

> *Now what does the public want in a post-office? Probably the first desiderata are pens that can be used for writing out telegrams... Another very common deficiency in post-offices is an adequate counter space and staff for the dispatch of telegrams. We must all of us have noticed how often there is a 'queue' of people wanting to send telegrams, which one harassed counter clerk is accepting, while there are yards of counter and employees doing nothing.*[4]

Because the Post Office had a monopoly on the services it offered there was no incentive to make post office interiors attractive to the public. However, attitudes within the Office of Works and the Post Office gradually changed. The Treasury sanctioned an increase in lighting levels and in 1929 the Post Office initiated a 'Brighter Post Offices' campaign: an attempt to improve its image and raise its

Right: Indian laurel-wood used for table and chairs at Birmingham head post office, 1930s.

Opposite, top: Public office in Streatham post office, 1929; a typical interior prior to the 'Brighter Post Office' campaign.

prestige by making the public office a more pleasant place to visit and work in.

Perhaps inspired by the decade's 'bright young things' and vogue for all things 'bright', the Post Office 'recognised its duty to provide seemly and commodious offices for the use of the public'.[5] Brighter schemes for mural decoration, better furniture fittings, more handsome counter screens, increased lighting levels, and more attractive flooring were introduced. After a few years it was claimed that the majority of Crown Offices presented a very creditable appearance. Other improvements included the introduction of writing tables and chairs in 1931 (a reversal of previous policy) and the installation of new wooden telephone cabinets.

The Office of Works policy was to specify timber from the Empire. During this period many new and unfamiliar timbers were introduced, with full use made of this choice. Birmingham head post office was refitted using Indian laurel. At other offices, timbers used included Australian walnut (Cambridge), English oak (Oxford), teak (Torquay and Sheffield), Honduras mahogany (St Austell), Burmese mahogany (Wallingford), Victorian oak (Herne Bay), West African walnut (Guildford) and Indian silver grey wood (Stockport).

Henry Seccombe's reconstruction of Plymouth head post office interior in 1933 was viewed as the forefront of the 'brighter' type of post office. Its decorations were designed to be in keeping with the ancient traditions of the city, with ornamental woodwork of British oak, and carvings of coats of arms associated with the city. The floor was particularly interesting: a green and white mosaic representing Plymouth's maritime history with a Mariner's Compass, an Elizabethan galleon and other early ships. The entrance lobby floor featured Drake's ship and the 'Mayflower' in mosaic. Yet the interior still incorporated many features (bronze mesh screens, pendant lighting, heavy furniture), soon to become outdated.

The 1937 *Head Postmasters Manual* outlined staff responsibilities with regard to the premises and equipment under their control. The manual was extremely detailed in the specification of equipment for the staff side of the counter, but less forthcoming in respect of the

Below, middle: Oil paintings behind the counter at Henley-on-Thames post office. Retention of the paintings was a condition of the sale of the building to the Post Office.

Below, bottom: Exeter head post office, 1933; an example of the use of wood panelling and mosaic flooring.

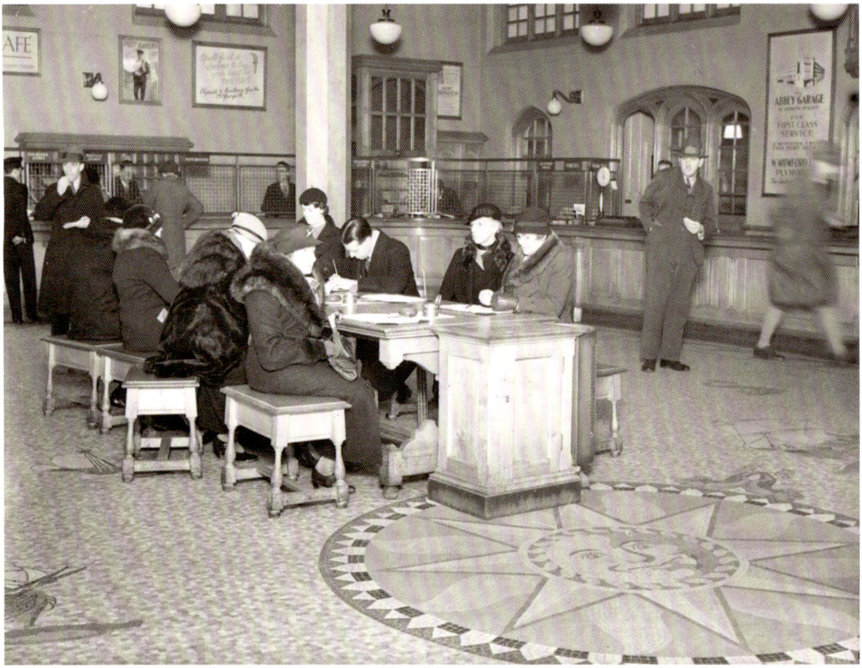
ABBEY GARAGE
FOR
FIRST CLASS
SERVICE

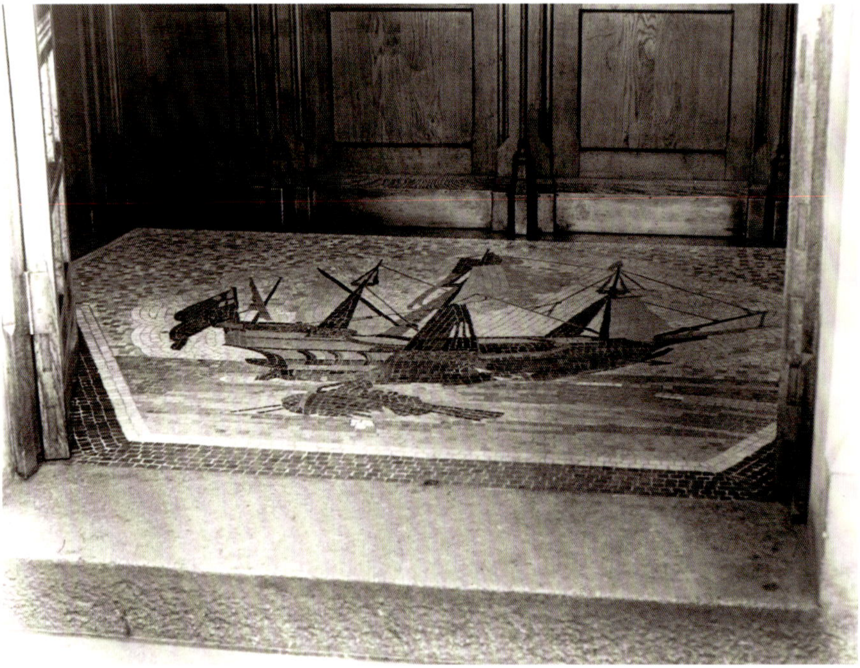

Opposite: The interior of Henry Seccombe's Plymouth head post office was regarded as a key example of what a 'brighter' post office should look like, the entrance to the public office included elaborate mosaic flooring celebrating Plymouth's heritage.

public side. It stressed that 'special attention' be given 'to the neat arrangement of notices, to the withdrawal of obsolete notices, and to the prompt replacement of any notices which become torn or dirty'.[6] Highlighting just how many services the Post Office provided, the *Manual* listed no less than forty-five standard counter business notices which could be displayed.

Two types of display posters were permitted: 'prestige notices' to draw the public's attention to Post Office work, and 'selling notices' used to promote Post Office activities, for example, the introduction of new services. In certain offices framed illustrations of ancient monuments were also displayed by arrangement with the Office of Works.

In terms of internal appearance, the manual highlighted a colour scheme chosen by the architects of the Office of Works; bronze grille counter screens; counter tops finished in green or blue linoleum; counter fronts of a simple design, counter notices in bronze frames; floors of ceramic or terrazzo tiles in simple geometric patterns (with symbolic designs provided only in exceptional cases); plinths of polished granite or other polished or glazed material; light fittings of the enclosed pendant-type; clocks of circular or octagonal design (with special designs provided only in the most exceptional circumstances); telegram writing tables, chairs and date indicators made of oak, mahogany or teak; and glass ash trays.

At some sites, there were exceptions to the standard decor. The walls of Henley-on-Thames post office were hung, not with the customary official notices and posters, but with a series of seventeenth-century oil paintings on wooden panels. These were originally housed in an Elizabethan mansion. When this was demolished in the 1830s the panels were removed to another house subsequently sold to the Post Office, a condition of the sale being that the panels be hung in the new post office erected on the site. At Woodbridge, Suffolk, the post office was a sympathetic conversion of an historic building and featured an elaborate plaster ceiling and chandelier lighting.

A crucial step in the 'Brighter Post Offices' campaign was the search for an alternative to bronze counter screens. The Post Office

Above: Colourful advertising posters became a key feature of post offices from the 1930s. At Plymouth, an example of a 'post early' campaign poster similar to one later designed by Graham Sutherland (above) can be seen (top, left) behind the post office counter.

Overleaf: Kings Lynn post office interior, c.1936, a fine example of a more modern, 'brighter' public office incorporating two telegram tables.

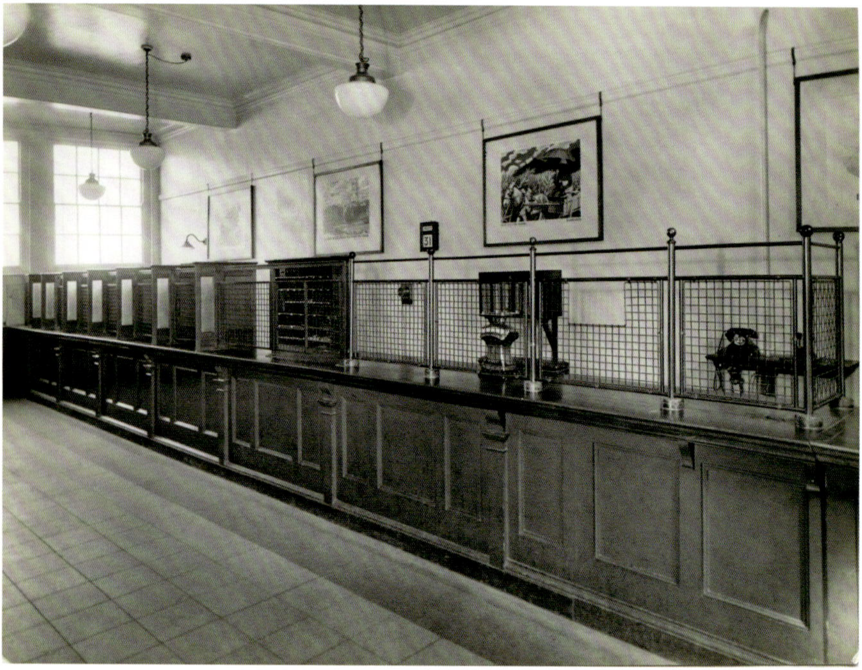

Opposite: Chelmsford post office counter with bronze grille, and later replaced with experimental Perspex counter screens following design trials in 1947.

Left: A busy 1930s post office interior, highlighting the growing need for interior design which facilitated quick, efficient service.

architect, W.H. Ludlow, initiated a series of experiments at a number of smaller offices between 1936 and 1939 with screens consisting of four glass louvres. There was a favourable reception. These screens were thought to be more hygienic, and of bright and attractive appearance, making it easier for staff and public to communicate with each other. Subsequently, the Office of Works was requested to install glass screens at new offices and at existing offices where it was desired to replace old-fashioned grilles. The outbreak of war in 1939, however, brought a halt to further progress.

The accepted method of staffing post office counters, by having different services available at different positions, showed its deficiencies during World War Two, when the number of customers increased with many people wishing to transact more than one item of business. This necessitated a visit to several separate positions. Because many female customers were engaged in wartime occupations they could visit the post office only at certain periods of the day. As such, they could not afford to spend too much of, perhaps their lunch hour, in the post office when other duties, such as shopping, had to be done in the same short space of time. An attempt was made to overcome this by introducing the Team Scheme. This allowed a group of counter officers to engage in several types of business, each of which had hitherto been the responsibility of individual members of staff.

In the immediate post-war years development of post office interiors ceased due to labour and materials shortages. By the early 1950s, however, the Post Office was again considering improvements. It was a temporary post-office, at the 1951 Festival of Britain on London's South Bank, which provided inspiration for a review of public office interiors. Designed by Michael Grice of the Architects' Co-Operative Partnership, its innovative layout of service points aroused considerable interest:

> *The public Post Office is... unusual in many respects. Entrance is through double plate-glass doors and natural lighting and ventilation are provided by twenty-four circular windows in the ceiling. The eleven position counter is of a new curved type and its grilles are of glass with vertical slits for speech. Each serving position is offset at an*

Opposite: The circular temporary post office at the British Empire Exhibition, Glasgow in 1938 (designed by Thomas Tait) featured a recessed lighting system, an approach that would not be repeated in Crown post offices until the mid 1950s.

Left and below: The temporary post office at the Festival of Britain exhibition, 1951 featured both a unique counter arrangement, which was later tested by the Post Office at a number of provincial sites, together with contemporary light fittings and redesigned writing tables.

Opposite: Littlehampton (top) and Scarborough (bottom) were two of the post offices chosen in the early 1950s for counter configuration experiments. Cost considerations prevented any further rolling-out of these schemes.

> angle of 20 degrees giving the appearance of a saw-tooth edge. It is claimed that this layout enables a larger number of customers to be served quicker than is possible with the present standard counter arrangement.[7]

Also in 1951, a report, *The Public Office: Some Observations on its Design, Fittings and Furniture* by John Evans of the Post Office's Northern Region, argued that a more appropriately designed office would do much to overcome the negative experience of using the post office. Evans suggested it was essential that public and staff be brought into closer contact, with obstacles to quick and efficient service removed. This would give staff a greater feeling of personal responsibility and spirit of service, while also creating a more pleasant environment for customers. Evans advocated that glass counter screens replace the bronze mesh screen which created a defensive 'them and us' culture. The 'saw-tooth' counter demonstrated at the Festival of Britain was his preferred solution.[8]

The Post Office subsequently experimented with the echelon or 'saw-tooth' counter, and with a variety of louvre and aperture glass counter screens at various Class II offices. Despite the advantages (particularly in long and narrow offices with a single entrance), the cost was prohibitive compared with the traditional straight counter and little further progress was sanctioned.

Nevertheless the episode demonstrated that the Post Office was willing to consider suggestions about the customer experience, and indirectly led to the creation of the 1953 *Working Party of the Design and Layout of Public Offices*. Its terms of reference were to:

> review existing rules, standards and practices governing the design, layout and lighting of public Post Offices and of the counters and other fittings contained within (excepting any matters which are exclusively the responsibility of the Ministry of Works), and to make any recommendations for changes and further standardisation which may seem necessary.[9]

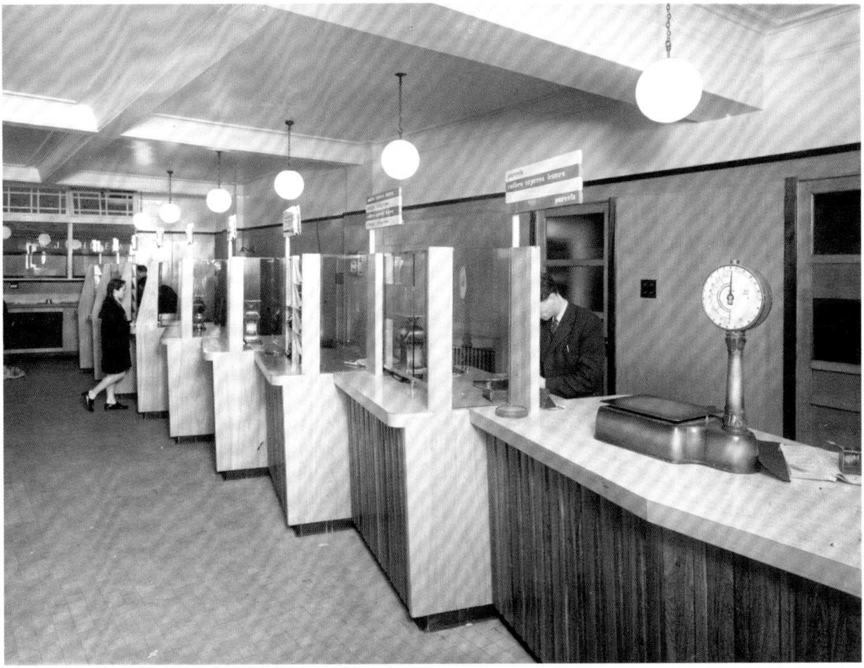

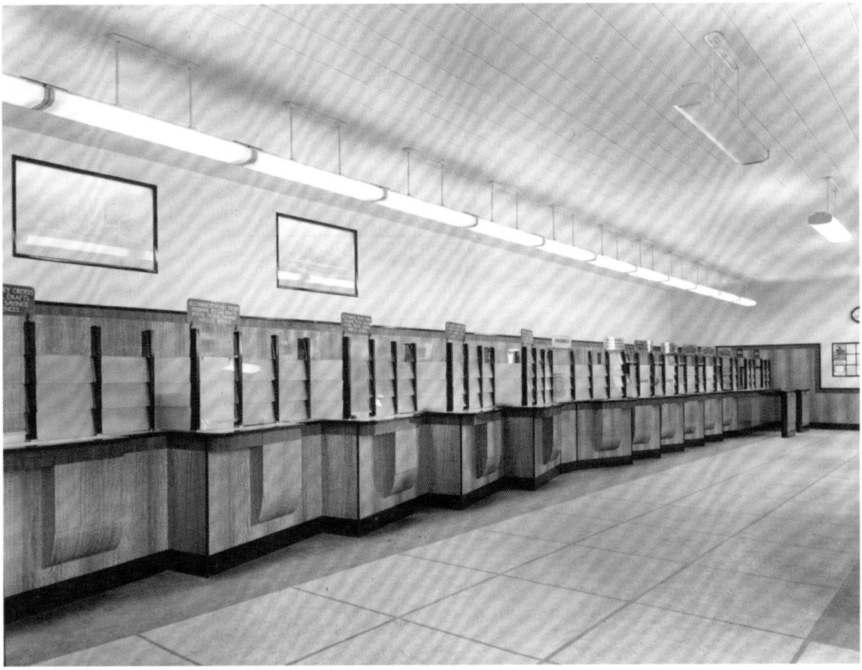

Over the next five years three reports were issued, each the result of exhaustive investigations into all aspects of the public office, including medical and audiology studies. For example, a study by the Treasury's Deputy Medical Officer concluded that members of staff working in offices with glass counter screens took fewer days sick leave than their colleagues behind the traditional bronze mesh screen. It was also discovered that the public found it difficult to position themselves in front of an aperture within a solid screen. A return to grille-less counters was even considered.

Eventually, the Working Party's first report (April 1956) recommended (twenty years after the first experiments had been conducted), that counter screens should be of louvred glass rather than bronze grilles, to dispel the public impression of the Post Office as a 'stand-offish, old-fashioned and unimaginative institution'. By the mid-1950s glass had also become a cheaper option to specify than bronze. The report also recommended that to achieve a brighter post office, floors should be chosen with regard to the locality, be hard-wearing, easily cleaned, and of pleasing appearance; walls should be plastered allowing a choice of paint colours to be specified by the architect, and ceilings should incorporate sound absorbing materials. As to the provision of writing tables, the Working Party concluded that, given the high cost of obtaining sites and providing buildings, interior space should be used in the most economical manner. Tables with chairs were not recommended, especially since it was

Opposite: The public area at Thomas Winterburn's post office at Berkhamstead; a fine example of interiors of the later 1950s.

Left: The Remnant Street branch office, Kingsway, London following refitting according to the recommendations of the Working Party on Public Offices in 1958.

known the facility would be abused by members of the public sitting at them for long periods of time. Telephone call boxes should no longer be situated inside the public office, but provided in an adjacent area, easily accessible from the public office.

The Working Party's second report (November 1957) gave further encouragement to architects to produce bold and imaginative designs (assuming no additional cost was incurred), and recommended that the approved glass screen be fitted on a standard counter. The provocative POSITION CLOSED signs, deeply unpopular and criticised by the press, were to be abolished. Any member of staff working on non-public duties should do so out of sight of the public. Illuminated business notices should be provided in larger offices, and 'flag' type notices in smaller ones. These could be switched off or removed when the position was unstaffed.

The Working Party's third and final report (June 1958) confirmed the four-louvre glass screen as standard, and recommended that counter surfaces not be limited to dark green and blue; linoleum was the preferred material. As for the dimensions of public office interior, all offices with counters up to 30ft long should have depths of 15ft in front of the counter, counters up to 20ft long should have depth of 12ft; the height of public offices should be reduced to 12ft if the counter length was up to 20ft, to 14ft if the length of counter was over 20ft.

The Remnant Street branch office, Kingsway, London was chosen to implement the Working Party's recommendations. Opened in 1958, it proved a great success. Frequent expressions of pleasure from members of the public entering the office were reported. Thanks to the efforts of Ministry of Works' architect Thomas Winterburn, the Kingsway office included new business notice signage in suspended illuminated boxes, with contemporary design lettering.

Cost considerations, however, prevented further developments, and it was recommended that business notice signage be incorporated in a cheaper illuminated trough along counter tops. Winterburn believed a contemporary post office should create a feeling of spaciousness through good design, eliminating fussy features like traditional pendant light fittings; chairs and tables should be replaced by writing tables

similar to those in the Festival of Britain post office. Winterburn further advised against providing shelves for umbrellas and handbags unless the Post Office wished to add lost property to its services!

In terms of decorative finish, the average post-war post office had a terrazzo floor in the public office – a hard wearing material available in a wide range of colours and patterns – replacing the mosaic floors used before the war. To achieve a modern look, marble ceased to be used. Plaster and paint were considered the most appropriate wall finishes, although plywood or blockboard veneer were also employed. No definite guidelines, however, were issued: the architect was free to specify whatever materials he thought appropriate assuming they could be justified economically.

In 1960 another enquiry into post office services was published. The *Joint Committee on Service at Crown Post Office Counters* (chaired by L.J. Taylor), in responding to complaints about queueing, closed positions, lack of courtesy and inadequate premises, recommended that Crown post offices should be warm, cheerful, well-ventilated, and properly lit, with a pleasant scheme of decoration. It recommended the introduction of composite working, enabling customers to do nearly all kinds of business at any position, rendering business notices redundant. It suggested that the closed position problem be solved by introducing a traffic light system: green for open counter positions, amber for one about to close and red for no service. Finally, after all the experimentation with optimum glass screens, the Committee concluded that the four-louvre glass screen was not satisfactory because it impeded hearing in noisy offices. Nevertheless, security reasons prevented the abolition of counter screens altogether.

Opposite: Croydon post office interior c.1970 illustrating modern lighting and flooring and counter positions no longer requiring individual business signs.

Right: Throughout the twentieth century a range of modular vending machines for stamps, stationery, and eventually posting, were trialled at post offices, introduced to reduce queuing and improve efficiency.

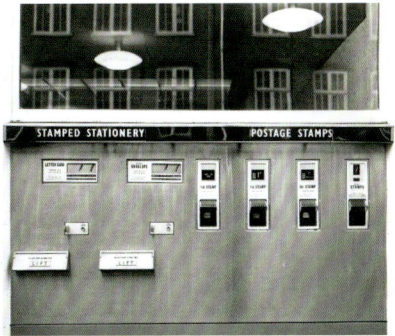

The Post Office modernisation exercise involving Hugh Casson, Misha Black, W. L. Stevenson and F. J. M. Ormrod also led to significant interior improvements. Ormrod and Stevenson's contribution was Corn Exchange, Liverpool, post office. Here they tried to recapture 'the tradition and even romance of the coaching days', using the 'heraldic colour' of bright red, yellow and blue, thereby transforming a post office of previously dingy appearance into a light and bright space. Traditional Post Office red (the standard colour since 1799) was replaced with a vibrant 'singing red', containing more orange. David Jones, the architect in charge of the modernisation declared 'old Post Office Red, used with black and other colours, is dead. We have used the orange-red with black or white or stainless steel. These bring the red out; they "send it" to use the modern idiom'.[10]

So attractive did the Corn Exchange office upgrade prove that it was called 'the coffee bar' by local people. At the Manchester Royal Exchange office additional colour was added to the interior by the linoleum floors and glass-fibre chairs. These post office interiors, along with those designed by Casson and Black, were lively, colourful and well-received, but very expensive. The challenge was to incorporate the best of these improvements in a more economical framework.

Hitchin head post office (1962) incorporated the new all-purpose counter service regime, recommended by the Taylor report, and a glass screen, not in the four-louvre format, but with three vertical apertures. It also featured a traffic light system indicating the status of each counter position. Although the building's external appearance was strictly functional, the imaginative use of materials and colour schemes broke away from what many expected on entering a post office. It also introduced a built-in system of modular vending machines and stationery designed by Douglas Scott (1913-99) with Associated Automation, subsequently known as the 'Hitchin Unit'.

Reporting on the 1962 post office exhibition at which the designs were displayed, the *Post Office Magazine* declared:

Too many post offices counters still look dark and antiquated, with heavy décor of polished mahogany, bronze and 'marble'. They don't suggest the modern, thriving business which is the Post Office

Right: Refurbished interior of Liverpool head post office, c.1980; transformed from its previously dark, early twentieth-century interior.

Opposite: Newport post office counter 1987, illustrating the ongoing impact of earlier post office interior experiments with its lighter interior colour scheme, louvre-glass counter positions and noise absorbent ceiling tiles.

today... Bronze grills are replaced by glass, old-fashioned dangling lamps disappear into the ceiling or at least acquire a modern look. Dark brown, bronze and cream are no longer the only possible colours... The intention is to have all 1800 main post offices smartened up by about 1968.[11]

The exhibition pointed towards a bright future for post office interiors. Internal walls would be lined with flush wood paneling, with no plasterwork at all; ceilings would be standard in all new offices – acoustic tiles with recessed fluorescent fittings. Standard treatment would be applied to the counter front, with wood paneling the preferred choice. In addition, typography would be standardised throughout – a discernable Post Office 'brand' would emerge. It was intended that these improvements, part of the new 'house-style', would be adopted in all new post offices. The Postmaster General hoped the improvements would make post offices more pleasant places to do business and work in, be achieved in a relatively short time, and at an economic cost.

Ultimately, as with the external standardisation of post offices, these hopes were not fully realised and the 'house-style' was eventually abandoned, Nevertheless, many ideas suggested by the exhibition and earlier experimental office designs lived on, reappearing in later post offices including those designed by private sector architects. Glass screens increasingly became the norm, position signs became minimal, and a colour red-based post office branding developed and evolved, uniting the UK's varied network of post office buildings into a recognisable, unified service.

Notes

[1] *Western Figaro Supplement*, Dec. 5, 1884.
[2] *Aberdeen Daily Journal*, 6 Apr. 1907.
[3] *Blackpool Gazette*, 8 Nov. 1910.
[4] Wellesley, Gerald 'Recent Post Office Architecture'. *Architects' Journal*, 6 Jan. 1926.
[5] *The Post Office: a Review of the Activities of the Post Office*. 1934.
[6] *Head Postmaster's Manual*. 1937.
[7] London, E.C. 'The Post Office and the Festival.' *Post Office Magazine*. May 1951.
[8] Evans, John *The Public Office: Some Observations on its Design, Fittings and Furniture*. Unpublished 1951.
[9] *Working Party of the Design and Layout of Public Offices*. 1953.
[10] 'Singing Red' comes out with a 'Bing'. *Post Office Magazine*. Jan 1961.
[11] 'Out of the Bronze Age.' *Post Office Magazine*, Sept. 1962.

4

An Architectural Legacy

POST OFFICE BUILDINGS TODAY

Changing economic circumstances and priorities have over time led to the closure of many older purpose-built Crown post offices. Most are from the Victorian and Edwardian eras, although a number of 'Post Office Georgian' buildings also now serve other purposes. The new uses to which these old post offices have been put are many and varied, embracing cultural, business, leisure and domestic uses. Many towns have more than one former post office building now meeting very different public needs.

This is not a new phenomenon, even in Victorian times post office buildings, often taken on a lease for twenty or thirty years, would be outgrown as the business offered more and more services to customers. As a result, the post office would move to another building, leaving its former home to be put to another use, or demolished to make way for another development. In some cases the changing shape of a town or city has led to the post office no longer occupying the most convenient location. Southampton is a good example; here the traditional high street has been overshadowed by a new shopping centre nearby. As a result, the post office ceased to be centrally located or in the most convenient place for customers, and consequently moved partly to respond to this changing topography.

In most cases, Victorian and Edwardian post office buildings have been afforded a measure of protection from demolition because they have been listed as buildings of historical and architectural importance. Representative examples of inter-war post office buildings are similarly protected – among their number are Henry Seccombe's offices at Glastonbury and Sidmouth, and H.T. Rees's office at Horncastle. The reality of this protection means that when a listed post office building is redeveloped the façade is retained, while the interiors are removed.

In the cultural sector attempts to transform larger post office buildings have not always been successful: these projects are dependent on public funding (either from government agencies or the National Lottery) and this has not always been forthcoming. In the 1990s plans were announced to convert the empty Glasgow head post office building into a new National Gallery of Scottish Art and Design. A design competition was held (won by Page & Park

Opposite: John Rutherford's Twickenham post office, built 1908, is still an impressive presence on the high street and is now a Wetherspoon's public house.

Left: The Crowndale Centre in Camden is housed in the former North Western District Office.

Opposite: Henry Tanner's Wolverhampton post office building is today part of the University of Wolverhampton.

Architects), but the project was abandoned after the Heritage Lottery Fund rejected its multi-million pound funding bid. In the provinces similar cultural transformations have been attempted. In 2005 plans were announced to convert Henry Tanner's Preston head post office into a library and knowledge centre in the heart of the city's newly designated cultural quarter. This would have provided a visitor attraction, housing important collections previously unavailable to the public, and a tourist information centre, acting as a hub and gateway for the city, county and region. The façade of the post office was to be retained and an environmentally-friendly glass building constructed inside and above the building offering views over the city, thereby regenerating a Grade II listed landmark. Sufficient funding was not however forthcoming to allow the project to proceed. Plans were then made in 2007 to convert the building into a business centre for young entrepreneurs. The building currently houses the gallery of Preston Art & Design, supporting and promoting the work of North West based artists and designers.

The head post office in Forster Square, Bradford, was built on land belonging to Bradford Cathedral. Following its closure after over a century of service, the Cathedral authorities used a National Lottery grant to develop the building (renamed St Peter's House) into a museum of religious life, celebrating the plurality of British faith. *Lifeforce*, as it was known, opened in July 2000, but was unable to sustain visitor numbers, closing in February 2001. The building was subsequently acquired by Kala Sangam, an organisation promoting greater understanding and appreciation of the cultural traditions of South Asia.

However, successful arts venue occupations of post office buildings include the Post Modern Gallery in Swindon, the Workplace Gallery in Gateshead, the Red Box Gallery in James Williams's post office building in Newcastle-upon-Tyne, and the Old Post Office gallery in Burslem, Stoke-on-Trent.

In the field of education and welfare, former post office buildings house Oldham Local Studies and Archives Department, the extension to Epsom School of Art, and part of Wolverhampton University. London's North Western District Office in Camden was converted

between 1987 and 1989, and is now home to the Crowndale Centre, housing a library and health centre.

For private sector developers, large Victorian and Edwardian post office buildings offer a major attraction – they occupy valuable prime city centre sites, ideal for prestigious offices and high quality residential accommodation. In Glasgow, for example, following the abortive attempt of the National Design Gallery project, the post office building has been developed into upmarket residential accommodation and office space known as G1. The façade has been retained, and a four-storey glass roof extension added, affording views across the city and beyond. Similarly, in Edinburgh the general post office has undergone a radical transformation. The building ceased to be used as a post office in October 1995, and in 2003 conversion into office accommodation began. It was one of Europe's largest-ever façade retention projects, featuring a glass exterior wall built behind the façade, a central atrium, and a roof garden. After years of non-occupation, the building, now known as Waverley Gate, is part

occupied by Microsoft, and set to become the headquarters of Creative Scotland, formed from a merger between the Scottish Arts Council and Scottish Screen.

In the City of London, the St Martin's-le-Grand area was from 1829 the headquarters of the General Post Office. After the demolition of Smirke's General Post Office building in 1912, three large buildings remained housing postal and telegraph services: GPO North (the administrative headquarters), GPO West or the Central Telegraph Office, and King Edward Building. The Post Office no longer retains a presence in this area. King Edward Building has been redeveloped as the UK headquarters of American banking group Merrill-Lynch; GPO North is now occupied by Japanese bank, Nomura. On the site of the Central Telegraph Office (now demolished) is the headquarters of British Telecom. Of Smirke's General Post Office building, nothing remains, although fragments have survived. Visitors to the gardens at Hyde Hall, Hertfordshire will find the Ionic capitals used as flower

Opposite: The prominent high street location of John Rutherford's Richmond post office building has made it the ideal premises for retailer HMV.

Right: Only traces remain of Robert Smirke's General Post Office building; one such is an Ionic capital gracing the entrance to the Vestry House Museum in Walthamstow.

pots; another capital was removed to a site outside the Vestry Museum in Walthamstow Village, East London. The only reminders of the Post Office's association with St Martin's-le-Grand are a statue of Rowland Hill outside the former public office at King Edward Building and the small area of green space nearby, known as Postman's Park.

Retail and mixed use developments have also been created in former post offices. Two of the most notable are the Met Quarter in Liverpool, and the Mailbox in Birmingham. Liverpool's former head post office has had an interesting history. Built to Henry Tanner's design between and 1894 and 1899, the building was severely damaged during the May blitz in World War Two. Although saved from demolition, the upper floors were removed. When the Post Office vacated the premises, the interior was gutted and the building shell remained for many years awaiting development. The site was acquired by a developer in 2003 and transformed into Liverpool's first upmarket shopping centre, known as the Met Quarter. Even under reconstruction the building contributed to the cultural life of the city: the hoardings surrounding the site were used as an outdoor art gallery by students of John Moores University.

Opposite: In front of the redeveloped Royal Mail Sorting Office, The Mailbox, a large public space allows ease of access and the free movement of visitors, while to the rear a bridge enables access to Birmingham's network of canals.

The most ambitious and successful urban regeneration project has been the Mailbox in Birmingham, formerly a Royal Mail sorting office. The building, designed by R.H. Ousman of the Ministry of Public Building and Works, was the largest parcel and letter sorting office in the country, using all the latest mechanised techniques. It was also the largest building in the centre of Birmingham. A tunnel linked the site to New Street Station, enabling the delivery of mails directly to the office. After less than thirty years use, Royal Mail sold the building to a developer in 1998. Over the next ten years the structure was demolished but for the steel frame and the site transformed into the most prestigious mixed-use development in the city and beyond, comprising the local offices of the BBC, hotels, retail spaces, a health club and apartments.

On a smaller scale, plans exist to convert the post office in Fitzalan Square, Sheffield, into a mixed-use development of restaurants, offices and a hotel following its sale by Royal Mail in 2005. Archibald Bulloch's post office in Bath has also been redeveloped into a mixed-use space comprising apartments on the upper floors, retail units and a modern post office at ground level, with Bath Postal Museum in its basement.

Former city or town centre post offices have been converted into residential use at a number of sites, most notably in Leeds, where Henry Tanner's City Square post office now houses a number of luxury serviced apartments. The façade of the former post office in Southport also disguises a number of luxury apartments, and in Tunbridge Wells, a gated development, Post Office Square, has transformed the former post office building. Other notable examples of post office building conversions into residential use include City Exchange, Kingston-upon-Hull, Aberdeen, and the upper floors of Tanner's Southampton post office. Numerous examples of smaller urban post office buildings converted into residential accommodation also exist, for example in Barnsley, Batley, Isleworth and Shoreditch.

Although many provincial and suburban post office buildings are not large enough to be profitably developed into office accommodation, their central location and attractive structure means many have been converted into licensed premises, restaurants, or coffee shops. There are numerous examples of this type of conversion, with many

Right: New Century House in Aberdeen has been created within and around the old post office.

Opposite: Henry Tanner's 1903 Sunderland post office building is a paradigm of how a former post office can contribute to the regeneration of a city, while fulfilling a social role. Sunniside, the original business centre of Sunderland, has in recent years attracted substantial investment for urban renewal. As part of this regeneration Tanner's building has been converted into apartments by a local housing association.

of those converted to public houses bearing names that reflect the building's former use. Typical are *The Last Post* in Loughton and Southend, *The Postal Order* in Blackburn, Crystal Palace and Worcester and *The Penny Black* in Beeston, Bicester, and Northwich.

An interesting footnote to the story of converted post office buildings is the ongoing preservation of associated memorials and plaques. The Post Office owns the second largest number of war memorials in the UK after the Church; testimony to the number of postal workers who served during two World Wars, often as part of the Post Office's own battalion, the Post Office Rifles. A Post Office War Memorial to those who died in the First World War, designed by Tyson Smith, was unveiled in the public office at Liverpool head post office in June 1924. Following the office's closure the memorial was moved,

In some cases the names of public houses converted from post offices recall former post office use – for example, The Last Post in Beeston and Loughton (left); The Penny Black in Bicester (below), Northwich and Southend; and The Postal Order in Blackburn.

Because of their central location and relative ease of conversion, former post offices have provided attractive development opportunities for restaurants, bars and coffee shops. Examples include ASK in Northwood (below, right), Zizzi in Surbiton (right) and Pizza Express in Didsbury, Manchester (below, left).

conserved and re-instated in the Met Quarter. Southampton's association with the story of the Titanic was similarly recorded on a plaque in the post office, listing the five postal staff who died in the tragedy. Following the building's closure, the plaque was removed, and re-located to the Civic Centre.

Like their associated memorials which mark the public service of former postal workers, post office buildings are an ongoing testament to the public service enabled by their design. While some have taken on new uses and others continue performing this role, all highlight the contribution to the UK's built environment made by the purpose-built post office.

Below: War and other memorials in former post offices have been retained on redeveloped sites or moved to new locations. The Met Centre in Liverpool, built on the site of the former head post office, now houses the post office war memorial previously located inside the public office.

Opposite: Perhaps the most original use of a post office building – in 2008 twenty-two artists were commissioned to create public art works for the first Folkestone Triennial. As a result, David Dyke's post office building in Tontine Street became the site of Nathan Coley's illuminated installation, HEAVEN IS A PLACE WHERE NOTHING EVER HAPPENS.

Henry Tanner (left) and David Dyke (right) – the most prolific of all Office of Works post office architects, each responsible for 'golden ages' of post office design.

List of Known Works

BY PRINCIPAL POST OFFICE ARCHITECTS

The following is a chronological list of the most prolific of the post office architects and the buildings which are known to have been designed by them. It is not intended to be comprehensive. The dates are approximate, indicating either when plans for the building were prepared, or when the building opened to the public. Dates of each architect are given where known.

Robert Matheson (1808-1877)

Matheson joined the Edinburgh Office of Works as a junior clerk in 1828 and rose to the position of surveyor for the Board of Works in Scotland.

List of known works:
Aberdeen (1875)
Dundee (1863)
Edinburgh (1859)
Glasgow (1875-8)
Leith (1873)
Paisley (c.1863)
Perth (c.1863)

James Williams (1824-1892)

Williams joined the Office of Works at the age of twenty-four. In 1859 he was appointed the first surveyor for the erection of post offices, a post he held until his retirement in 1884. An obituary praised his 'conspicuous zeal and ability' in superintending this very important branch of public service.

List of known works:
Carlisle – Lowther Street (1863)
Hull – Market Place (1875)
London – Bedford Street, Covent Garden (1883-4)
　　　　　Lower Tooting (1884)
　　　　　North Eastern District Post Office, Bethnal Green (1860)
　　　　　GPO West / Central Telegraph Office (1874)
Manchester – Spring Gardens (1884)

Newcastle-upon-Tyne – St Nicholas Street (1871-4)
Nottingham (1868)
Scarborough (1881)
Stockton-on-Tees (1878-80)

Walter Wood Robertson (1845-1907)

Robertson joined the Office of Works in London in 1871. In 1877 he succeeded Robert Matheson as the principal architect and surveyor for Scotland. He retired from his post in 1904, to work in private practice.

List of known works:
Aberdeen (1907, with William Thomas Oldrieve)
Airdrie (1905)
Dundee (1898)
Edinburgh (1890 extension)
Falkirk (1893)
Fife (1889-90)
Galashiels (1894)
Glasgow (1892-4 extensions)
Greenock (1898-9)
Hamilton (c.1900)
Helensburgh (1893)
Linlithgow (1903)
Paisley (1893)
Stirling (1894)

Sir Henry Tanner (1849-1935)

The most well-known of all the Post Office architects, Tanner worked in the Office of Works as assistant surveyor from 1871 to 1876.

Following a period in the office of Sir John Taylor, he returned in 1881 to become for the next two years surveyor for the Northern District, before becoming principal architect (1884-9) and eventually chief architect to the Office of Works until his retirement in 1913. Subsequently he joined his son, Henry, in private practice. Tanner had a particular interest in the structural aspect of the building process and served as President of the Institution of Structural Engineers, then known as the Concrete Institute.

List of known works:
Bedford (1897)
Birmingham (1889-91)
Bradford (1886)
Brighton (1900 enlargement)
Bury St Edmunds (1895)
Cardiff (1896)
Croydon (1894)
East Grinstead (1896)
Grimsby (1882 and 1892 extensions)
Halifax (1885)
Keighley (1880)
Leeds (1892-96)
Leicester (1887)
Liverpool (1894-9)
London – GPO North (1889-95)
         King Edward Building (1910)
         Mount Pleasant (1889)
         North Western District Post Office
         (1893 extension)
         Paddington District Parcels Office
         (c.1900)
         South Kensington (1908-9)
         West Kensington (1903-4)

West Central District Post Office (1895-7)
Nottingham (1895-8)
Preston (1901-3)
South Shields (1890)
Southampton (1894)
Stamford (1896)
Sunderland (1903)
Tipton (1897)
West Hartlepool (1897-1900)
Wisbech (1887)
Wolverhampton (1895)
York (1884)

William Thomas Oldrieve (1853-1922)

The most prolific of all the Edwardian post office architects, Oldrieve was unique in that he built extensively in England, Wales and Scotland. His first major appointment within the Office of Works was as architect for provincial post offices in England and Wales from 1892 until 1904, when he gained the position of chief architect for Scotland until his retirement.

List of known works:
Aberdeen (1907, with Walter Wood Robertson)
Aldershot (1902)
Ayr (1907)
Banff (1906)
Barrow (1903 extension)
Barry (c.1903)
Bathgate – West Lothian (1913)
Bootle (1905)
Brechin (1910)
Broughty Ferry (1907)
Burnley (1903)
Burslem (1903)
Burton-on-Trent (1905)
Crewe (1903)
Crieff (1906)
Dumbarton (1908)
Edinburgh (1907-11 extensions)
Glasgow (1915-6 extensions)
Glasgow Parcels Office (1903-5)
Haddington (1908)
Hereford (1903 alterations and additions)
Hinckley (1902)
Hyde (1899)
Kelso (1910)
Kilmarnock (1907)
Leicester (1905 extension)
Lerwick (1909)
Lichfield (1904)
Lincoln (1903)
Manchester Parcels Office (1906)
Merthyr (1905)
Montrose (1907)
North Berwick (1906)
Norwich (1903-4 extension)
Oban (1908)
Peterborough (1904 alterations and additions)
Plymouth (1904 enlargement)
Portsmouth (1903 alterations and additions)
Stalybridge (1899-1900)
St Andrews (1906 alterations)
Stornaway (1907)
Swansea (1898-1901)
Weston-super-Mare (1900)
Wick (1912)
York (1902 extension)

Walter Pott (1864-1937)

Pott joined the Office of Works as an assistant surveyor in 1886. He was appointed architect for post offices for the North of England in 1905, eventually becoming one of the three principal architects at the Office of Works from 1912 until his retirement in 1921.

List of known works:
Bacup (1910)
Birkenhead (1907)
Bishop Auckland (1911)
Blackburn (1907)
Blackpool (1910)
Blyth (1911)
Dewsbury (1908)
Kingston-upon-Hull (1908)
Leeds – City Square (1911 alterations)
Hyde Park Delivery Office (1906)
Liverpool – Walton Road (1905)
Malton (1911)
Manchester – Newton Street (1907-8)
Newcastle-upon-Tyne – Heaton (1911)
                     St Nicholas Street
                     (1909 alterations
                     and additions)
Rotherham (1907)
Scarborough (1905-9)
Sheffield (1910 addition)
Warrington (1906)

John Rutherford

Along with William Oldrieve and Walter Pott, Rutherford was one of the principal post office architects of the Edwardian era, working mainly in the south of England.

List of known works:
Aldershot – North Camp (1905)
Boston (1906)
Canterbury (1906)
Cheltenham (1906)
Dorchester (1904-5)
Droitwich (1907)
Dudley (1908)
Grimsby (1908)
Ifracombe (1909)
Leek (1906)
London – Acton (1911)
         Battersea (1912 extension)
         Kingston Sorting Office (1907)
         Notting Hill (1910 additions)
         Poplar (1911)
         Richmond (1904)
         South Eastern District Office (1911)
         South Western District Office (1892)
         Sutton (1906)
         Twickenham (1907-8)
         Uxbridge (1908)
St Helier (1908)
Stoke-on-Trent (1904-6)
Taunton (1909-11)
Torquay (1910-12)
Walton-on-Thames (c.1908)
Weymouth (1905)

Jasper Wager

Wager worked almost exclusively in the London area, having joined the Office of Works in 1877 and passing the qualifying examinations in 1898. An Associate of the Royal Institute of British Architects, he resigned in 1920, presumably on his retirement.

List of known works:
Esher (1902)
High Wycombe (1900-1)
London – Aldgate East (1908)
        Barnet (1904)
        Chelsea (1905)
        Eastern District Office
        (1905 extension)
        Enfield (1906)
        Ilford (1903)
        Kentish Town Sorting Office (1903)
        Knightsbridge (1903)
        West Brompton (1900)
        Northern District Office (1906)
        North Western District Office
        (1903-5 extension)
        Walham Green, Fulham (c.1900)
        Western District Office (1908-9)
        Wimbledon
        (1890 and 1901 alterations)
        Woodford Green (1904)

Henry A. Collins

Collins joined the Office of Works in 1886.

List of known works:
Bicester (1914)
Eastbourne (1911)
Frome (1914)
Lowestoft (1909 extension)
Margate (1910)
Petworth (1910)
Ramsgate (1908)
Tunbridge Wells (1911 additions)

Edward Cropper

List of known works:
Devonport (1911 extension)
Eltham (1911)
Leamington Spa (1911 extension)
Lichfield – Whittington Barracks (1912)
London – Hammersmith (1919)
        Kew Gardens (c.1928)
        Teddington (1926)
        Threadneedle Street (1926?)
Mansfield (1912)
Newcastle-under-Lyme (1912)
Stafford (1912 extension)
Tidworth Camp, Wiltshire (1910)

127

Albert Robert Myers (d.1962)

Myers joined the Office of Works in 1899, passing the qualifying examination in 1905. As senior architect to the Office of Works, Myers was engaged to some to the largest post office building projects.

List of known works:
Bridlington (1927)
Bulford Camp, Wiltshire (1914)
Clacton (1920)
Dover – Priory Street (1913)
Gillingham, Dorset (1915)
Haywards Heath (1915)
Ingatestone (1915)
London – Barking (1914)
Faraday House Telephone Exchange (1932-3)
Isleworth (1919)
Mount Pleasant (1927 and 1937)
North Western District Office (1919 completion)
Romford (1911)
Wembley (1920)
Woolwich (1914 additions)
Northampton (1914)
Penarth (1936)
Petersfield (1920)
Portsmouth (1912 alterations and additions)
Reading (1925)
Rochester (1912)
Southampton (1915 extension)
St Austell (1920)
Tonbridge (1915)
Wembley (1922)
Weybridge (1912)

Charles P. Wilkinson

Wilkinson was a prolific architect in the Northern Region of the Office of Works before and after World War One.

List of known works:
Accrington (1920 extension)
Barrow in Furness – Abbey Road (1930)
Bingley (1913)
Birkenhead – Liscard (1912)
Blackburn (1924 extension)
Blackpool (1924 extension)
Blyth (1911)
Bolton (1918)
Carlisle – Warwick Road (1913)
Chorley (1924)
Clitheroe (1926)
Colne (1926)
Darlington (1912 enlargement)
Eccles (1930)
Formby (1922)
Halifax (1923 extension)
Hoylake (1924)
Huddersfield (1912)
Kendal (1928)
Lancaster (1920)
Manchester – Spring Gardens (1926 alterations)
Stretford (1928)
Maryport (1914)
Middlesbrough (1913 extension)
Northwich (1914-5)
Preston (1925 alterations)
Rawtenstall (1921)
Rochdale (1927)
Sale (1930)
Scunthorpe (1914)

Southport (1922 extension)
Sowerby Bridge (1922)
St Anne's-on-Sea (1921)
Stockport (1912 additions)
Todmorden (1924 extension)
Widnes (1924)
Workington (1913)

H. T. Rees

A contemporary of Charles Wilkinson, Rees worked in the Midland counties and the North of England.

List of known works:
Chesterfield (1924)
Durham – Claypath (1927)
Goole (1926 alterations and extensions)
Gosforth (1923)
Horncastle (1929-30)
Ilkeston (1920)
Leeds – City Square
    (1924 additions and alterations)
Loughborough (1928)
Louth (1928)
Middlesbrough (1926 extension)
Northallerton (1929)
Redcar (1928)
Retford (1921)
Sheffield – West Street (1927)
Shipley (1924)
Skegness (1928)
Wisbech (1930 extension)

Henry Edward Seccombe (b. 1879)

Seccombe joined the Office of Works in 1904, passing the qualifying examination in 1909. Like his colleague Archibald Bulloch, Seccombe worked on post offices in the West of England and Wales.

List of known works:
Brixham (c.1935)
Chipping Norton (1930)
Glastonbury (1938)
Lampeter (1933)
Maesteg (c.1935)
Malvern (1936)
Penn (c.1933)
Plymouth (1933 interior reconstruction)
Sidmouth – Radway (1938)
Welwyn Garden City (1930)
Westbury-on-Trim (c.1934)
Worcester Sorting Office (1935)
Yeovil (c.1932)

David Nicholas Dyke (b.1881)

The most prolific of the inter-war Office of Works architects, Dyke, in his application to become a Fellow of the Royal Institute of British Architects, claimed to have designed over a hundred buildings, although not all post offices. He joined the service in 1913 and remained there for the rest of his career. Most of his buildings were in the South of England.

List of known works:
Amersham (1928)
Ascot (c.1934)
Basingstoke (1926)
Beaconsfield (1925)
Beccles (c.1928)
Bedford (1924)
Bexhill-on-Sea (1931)
Billericay (1938)
Bognor (1924)
Braintree (1929)
Camberley (c.1936)
Cambridge (c.1935)
Chichester (1937)
Cosham (c.1927)
Coventry – Hertford Street (1923)
Crawley (c.1928)
Crowborough (1925)
Dartford (1925)
Dorchester (1919 extension)
Dorking (1930)
Edenbridge (c.1930)
Folkestone – Tontine Street (c.1937)
Gillingham (1924)
Gorleston (c.1937)
Gravesend (1925)
Harpenden (1927)
Hastings (1927)
Herne Bay (c.1935)
Hove (c.1927)
Kings Lynn (1939)
Leigh-on-Sea (1925)
Maidstone (1928)
Newbury (1929 extension)
Parkstone (1927)
Slough (1924 additions and alterations)
Staines (1930)
Thetford (1937)
Uckfield (1928)
Ware (1928)
Wareham (c.1930)
Winton (1924)
Wokingham (1929)
Woodbridge (1934)
Worthing (1928-30)

Archibald Bulloch (b.1882)

Bulloch passed the Office of Works examination for assistant architect in 1900, and after working in China between 1910 and 1915, was appointed district architect for Manchester between 1910 and 1915. The rest of his career was spent as architect for post offices in the West of England and Wales.

List of known works:
Aberdare (1924)
Abertillery (1924)
Ammanford (1925)
Avonmouth (c.1927)
Bath (1926)

Bodmin (1925)
Bridgend (1922)
Chepstow (1922)
Crewe (1927 alterations and extensions)
Egham (1928)
Evesham (1926)
Falmouth (1928)
Ludlow (1925)
Macclesfield (1925)
Maidenhead (1922 extension)
Nantwich (c.1928)
Neath (c.1928)
Sherborne (1928 extension)
Walsall (1927)
Wellington (1926)

Frederick A. Llewellyn

Llewellyn worked with his contemporary David Dyke on post offices in the South of England in the 1930s. Many of his buildings are in the London area.

List of known works:
Hatfield (c.1936)
London – Beckenham (1939)
          Dagenham (c.1932)
          Edgware (1929)
          Gerrard Street (c.1936)
          Greenford (c.1938)
          Hounslow (c.1934)
          Pinner (1930)
          Southall (c.1936)
          Stanmore (c.1933)
West Drayton (c.1935)
St Albans (c.1934)

Wallington (1930)
West Wickham (c.1930)

Archibald Scott

List of known works:
London – Harrow (1928 extension)
          Hackney (1927)
          Kentish Town (1930)
          Mitcham (c.1929)
          Norbury (1927)
          North Finchley (1927)
Loughton (c.1930)

Thomas Winterburn

Winterburn was chief architect to the Ministry of Works in the 1950s.

List of known works:
Berkhamstead (c.1958)
Canvey Island (c.1959)
Diss (1953)
Harlow – Stone Cross (c.1959)
London – Kingsway interior (1952)
Luton (c.1958)
Norwich (c.1956)
Saxmundham (1954)

Publications

Clarke, Jonathan. *Purpose Built Post Offices: A desk-based assessment of building type*, English Heritage, 2008.

Daunton, M. J. *Royal Mail: The Post Office since 1840*. London: Athlone, 1985.

Hemmeon, J.C. *A History of the British Post Office*. Cambridge: Harvard University, 1912.

Lewins, William. *Her Majesty's Mails: An historical and descriptive account of the British Post Office*. London, 1864. 2nd ed. 1865.

Reynolds, Mairead. *A History of the Irish Post Office*. Dublin: MabDonell Whyte, 1983.

Robinson, Howard. *Britain's Post Office: A history of development form the beginnings to the present day*. London: Oxford University Press, 1953.

Robinson, Howard. *The British Post Office: A history*. Princeton: Princeton University Press, 1948; Westport, Connecticut: Greenwood Press, 1970.

Stray, Julian. *Post Offices*. Oxford: Shire Publications, 2010.

Places to Visit

The British Postal Museum & Archive is the leading resource for British postal heritage. It cares for the visual, physical and written records of over 400 years of postal heritage including stamps, poster design, photography, staff records and vehicles. The BPMA is custodian of two significant collections: the Royal Mail Archive and the museum collection of the former National Postal Museum. The Royal Mail Archive is Designated as being of outstanding national importance. To find out more visit www.postalheritage.org.uk

The British Postal Museum & Archive operates at three sites:

Head office and Public Search Room for The Royal Mail Archive
Freeling House, Phoenix Place, London WC1X 0DL. Telephone 020 7239 2570. Public search room for archive items including Post Office records, photographs, stamp artwork, and staff records, based in Central London, open throughout the week, see website for details.

The British Postal Museum & Archive's Museum Store
Unit 7, Imprimo Park, Lenthall Road, Debden, Essex, IG10 3UF. Containing larger items including pillar boxes and vehicles, open on selected dates throughout the year, see website for details.

The British Postal Museum & Archive's Museum of the Post Office in the Community and replica Victorian Post Office, Blists Hill Victorian Village, Ironbridge Gorge Museums, Coalbrookdale, Telford, Shropshire, TF8 7DQ. For opening times visit www.ironbridge.org.uk

Index

Architects
  Allison, Sir Richard 50, 59; Bedford, Eric 75; Black, Misha 74, 105; Bulloch, Archibald 52, 55, 57, 115; Casson, Hugh 74, 105; Collins, Henry 47, 127; Cropper, Edward 127; Dumble, Alan 67; Dyke, David 50. 51, 52, 55, 59, 121, 122, 130; Elliott, Archibald 4; Grice, Michael 97; Hardwick, Philip 16; Howdill, Charles B. 20; Johnson, John 16; Johnston, Francis 3; Jones, David 105; Kay, Joseph 4; Llewellyn, Frederick 55, 59; Ludlow, W.H. 38, 97; Matheson, Robert 16, 123; Myers, Albert 59, 62, 63, 128; Nash John 4; O'Farrell, E.C. 74; Oldrieve, William Thomas 16, 27, 28, 125; Ormrod, F.J.M. 74, 105; Ousman, R.H. 115; Owen, William 16, 17; Page and Page 109; Palmer, Frederick 38, 42; Parr, John 74; Pinfold, Cyril 67; Pott, Walter 28, 47, 84, 126; Ralph, W.H. 67; Rees, H.T. 55, 59, 109, 129; Rivers, Edward George 14, 83; Robertson, Walter Wood 16, 23, 27, 83, 124; Rutherford, John 28, 32, 33, 109, 114, 126; Scott, Archibald 52; Seccombe, Henry 55, 57, 93, 109, 129; Shaw, Richard Norman 16; Smirke, Sir Robert 1, 4, 112, 113; Soane, Sir John 4; Stevenson, W.L. 74, 105; Tait, Thomas 99; Tanner, Sir Henry 20, 22-24, 50, 66, 83, 85, 110, 111, 113, 115, 116, 122, 124-125; Trevail, Sylvanus 16; Wager, Jasper 28, 33, 34, 43, 127; Watkinson, Philip 74; Wightwick, George 8; Wilkinson, Charles 47; 48, 59, 127-128; Williams, James 11-14; 16, 20, 123-124; Winterburn, Thomas 67, 70, 71, 74, 103
Architectural competitions 4, 16, 17, 20

Architectural styles
  Classical 4, 7, 8, 11, 14, 16, 20, 48, 52; Domestic Revival 47; Dutch/Jacobean 28; Edwardian Baroque 28; Flemish Renaissance 20, 23; Free Renaissance 28; French Renaissance 22-23; Greek Revival 3, 4; High Renaissance 4, 14, 16; Italianate 7, 11, 20; Medieval 48; Modernist 62, 63, 67, 70; Neo-Classical 47; Northern Renaissance 20; "Post Office Georgian" 47, 50-51, 52, 55, 59, 62, 63, 67, 70; Renaissance 11, 48, 55; Scottish Baronial 27, 28; Tudor Gothic 23, 27
Bath Postal Museum 115
Bevins, Reginald 75
Brick, preference for 48, 50
'Brighter Post Offices' campaign 89, 93
Cawley Committee on Post Office Buildings 40, 42, 43
City Exchange, Hull 115
City Square, Leeds 115
Clarke, W. 6
Coley, Nathan 121
Committee on Accommodation and Fittings 88
Committee on Post Office Fittings 85-86
Creative Scotland 112
Crowndale Centre, Camden Town 110
Epsom School of Art 110
Evans, John 100
Festival of Britain 67, 99, 104
Head Postmasters' Manual 90-91
Hennibique reinforced concrete system 23
Hill, Rowland 6, 113
'Hitchin Unit' 105
Hobhouse Select Committee on Post Office

Servants 24, 40, 43
'House-style' 75, 76, 80, 106
Howdill, Charles 2, 20
Joint Committee on Service at Post Office Counters 104
Kala Sangam, Bradford 110
Le Corbusier 67, 71
Lewins, William 6
Lifeforce, Bradford 110
Local materials, use of 40
Mailbox, Birmingham 115
Masey, Francis 20, 21
Merrill-Lynch Headquarters, London 112
Met Quarter, Liverpool 113, 121
Microsoft, Edinburgh 112
Ministry of Public Building and Works 74, 80, 115
Ministry of Works 66, 67, 71, 74
Modernist architecture, influence of 63-71
National Gallery of Scottish Art and Design, Glasgow 109
National Prize Medal Design for a Post Office 2, 20
National Telephone Company 40
New Century House, Aberdeen 116
Nomura Bank, London 112
Office of Works 3, 11, 14, 16, 20, 23, 27, 28, 32, 33, 34, 38, 40, 42, 43, 47, 48, 50, 51, 52, 55, 59, 62, 63, 66-67, 85, 88, 89, 90, 93, 97
Old Post Office Gallery, Burslem 110
Oldham Local Studies and Archives Department 110
Post Modern Gallery, Swindon 110
Post Office
  'Brighter Post Offices' campaign 89, 93; composite working schemes 104; lack of architectural expertise within 9, 38; leasehold arrangements 8, 9; private sector architects, employment of 3, 74, 80; staff appointments sanctioned by the Treasury 9, 38; survey of premises 9; travelling surveyor, request for 9; work to be approved by 20
Post office branches
  Aberdare 55; Aberdeen 9, 16, 27, 28, 115, 116; Accrington 47; Aldershot 28; Banff 28; Barnet 33; Barnsley 115; Barrow-in-Furness 28; Barry 28; Bath 9, 55, 115; Batley 115; Beccles 52; Beckenham 63, 66, 67; Bedford Street, Covent Garden 14; Beeston 117; Belfast 9; Berkhamstead 74, 103; Bexhill-on-Sea 52; Bicester 117, 118; Birkenhead 9, 28;

Birmingham 6, 22, 23, 90; Bishop Auckland 32; Blackburn 28, 13, 117; Blackpool 32, 84; Bodmin 55; Bolton 48, 85; Bootle 28; Boston 32; Bradford 20, 22, 110; Braintree 52; Bridlington 59; Bristol 9, 11, 14; Brixham 55; Burnley 28; Burslem 18, 110; Burton-on-Trent 28; Cambridge 52, 55; Canterbury 32; Canvey Island 74; Cardiff 22, 24; Central Telegraph Office 11, 13-14, 112; Chelsea 33; Chelmsford 97; Cheltenham 32; Chesterfield 59; Chichester 52; Chipping Norton 55; Coalville 38; Cosham 52; Cranleigh 67; Crewe 28; Crieff 28; Croydon 105; Crystal Palace 117; Dagenham 59; Derby 11; Devonport 7, 9; Dewsbury 28, 32; Didsbury 119; Diss 70; Doncaster 8; Dorchester 32, 33; Dover 59; Dublin 3, 4; Dundee 16, 23; Edenbridge 52; Edinburgh 3, 4, 5, 11, 16, 111; Enfield 33; Epsom 110; Evesham 52, 57; Exeter 17, 66, 67, 91; Faversham 71; Festival of Britain 97, 99; Folkestone 121; Galashiels 23, 26; Gateshead 110; General Post Office 1, 3, 4-5, 14, 112, 113; Glasgow 9, 16, 109, 111; Glasgow Empire Exhibition 99; Glastonbury 55, 57, 109; Gosforth 59; GPO North 22, 112; Greenford 59; Grimsby 22, 32, 33; Guildford 80; Haddington 28; Halifax 22, 47; Hanley 32; Harlow New Town 74; Harrow 76; Hatfield 59; Haywards Heath 59; Henley-on-Thames 91; Hereford 14, 28; Herne Bay 50, 51, 59; Hinckley 28; Hitchin 71, 76, 105; Horncastle 109; Hounslow 59; Huddersfield 48; Hull 9, 11, 28, 115; Hyde 28; Ilkeston 59; Ingatestone 38; Inverness 9; Isleworth 115; Islington 33, 34, 43; Keighley 22; Keynsham 67; Kilmarnock 28; King Edward Building 22, 23, 24, 50, 63, 83, 112; King's Lynn 93; Kingswood 74, 76; Knightsbridge 74, 76, 79; Knutsford 16, 17; Lampeter 55; Lancing 70; Leamington Spa 6; Leeds (Chapeltown Road) 74; Leeds (City Square) 9, 20, 22, 38, 115; Leicester 7, 28; Leigh-on-Sea 51; Leith 16; Lerwick 28; Lichfield 28; Lincoln 28; Littlehampton 100; Liverpool 9, 22, 88, 106, 117, 121; Liverpool (Corn Exchange) 74, 105; Liverpool (Eastern District) 22; Lombard Street 3; Loughton 52, 117, 118; Louth 59; Ludgate Circus 74; Luton 67, 71; Maesteg 55; 57; Maidstone 52, 55;

135

Malton 47; Malvern 55; Manchester 9, 10, 48, 66; Manchester (Royal Exchange) 74, 105; Margate 47; Merthyr Tydfil 28; Middlesbrough 47, 59; Minehead 42; Musselburgh 28; Mount Pleasant 62; Muswell Hill 63; Newcastle-upon-Tyne 11, 110; Newmarket 67; Newport 106; North Berwick 28; North Western District Office 110; Northallerton 59; Northampton 11; Northwich 48, 117; Northwood 119; Norwich 16; Nottingham 6, 7, 9, 22; Oban 27; Oldham 110; Oxford 14, 90; Paisley 16, 26; Parkstone 52; Penarth 63, 67; Perth 16; Peterborough 28; Plymouth 14, 66, 67, 83, 90, 93; Portsmouth 28; Preston 22, 47, 110; Pudsey 74; Reading 63; Redcar 59; Remnant Street 103; Richmond 113; Rochdale 43; Rochester 59; Rotherham 28; Saxmundham 74; Scarborough 28, 100; Scunthorpe 47, 63; Sheffield 8, 43, 115; Shipley 59; Shoreditch 115; Sidmouth 55, 57, 109; Skegness 59; South Kensington 22; South Molton Street 74, 79; Southampton 16, 17, 109; Southend 117; Southport 47; St Austell 90; St Helier 32, 33; Stockport 47, 90; Stockton-on-Tees 11, 74; Streatham 90; Surbiton 119; Sunderland 22, 116; Swansea 28; Swindon 110; Taunton 32; Tonbridge 59; Torquay 32, 90; Trafalgar Square 74; Truro 16, 59; Uckfield 52; Wakefield 11; Wallingford 90; Ware 55; Warrington 28; Welwyn Garden City 55; Wembley 62; West Drayton 59; West Kensington 22; Weston-super-Mare 28, 40; Weybridge 59; Weymouth 32, 40; Wolverhampton 11, 22, 110; Woodbridge 52, 93; Woodford Green 33, 34, 67; Worcester 117; Yeovil 55; York 22

Post Office Square, Tunbridge Wells 115

Post offices
    Class I offices 11, 20, 34, 38, 40, 42, 51, 85; Class II offices 20, 34, 38, 40, 42, 100; Class III offices 42; drive-in 74; modular vending machines 105; new uses 115-117; opening ceremonies 7, 14, 59; poor condition 16; public health issues 17; standardized design 71, 106

Postmaster General 1, 3, 5, 20, 32, 34, 38, 40, 43, 47, 75, 106

Prefabrication 71

Property Services Agency 22

Public houses 109, 118

Public office
    business notices 93, 103; colour schemes 89, 93, 102, 104, 105, 106; counter design 83, 84, 85, 88, 93, 97, 99, 100, 103, 106; counter screens 84, 85, 86, 87, 90, 93, 97, 100, 102, 103, 104, 105, 106; Empire timber specified 90; interiors 82-106; lighting levels 89; position closed notices 103; poster displays 93; queue control systems 104, 105; seating provision 85, 88; telephone boxes 90, 103; writing tables 83, 84, 85, 88, 90, 93, 99, 102, 104

Receiving houses 5

Reconstruction Committee for Emergency Building 48

Red Box Gallery, Newcastle upon Tyne 110

Restaurants 119

Royal Academy 59

Royal Institute of British Architects 20

Samuel, Herbert 32, 42, 47

Scott, Douglas 105

Shand, P. Morton 52

Smith, Tyson 117

Sorting offices
    Birmingham 115; Clapham 33; Dulwich 33; Finsbury Park 33; Hanwell 33; Kentish Town 33; Leytonstone 34; Lower Edmonton 34; 35; Mount Pleasant 62; New Southgate 34; Norwich 71; Tooting 34; Walthamstow 34; Whetstone 34; Winchmore Hill 34, 35

St Martin's-le Grand 1, 4, 112

Sutherland, Graham 93

Taylor, L.J. 104

Telephone exchanges
    Altrincham 71; Faraday House 62; Leicester 74; Leigh-on-Sea 50

Tite Prize Design for a Post Office 20

Townroe, B.S. 50

Treasury 1, 4, 9, 11, 17, 34, 38, 40, 43, 85, 89, 102

University of Wolverhampton 110

Vestry Museum, Walthamstow 113

War memorials 117-118

Wellesley, Gerald 50, 89

West Country Mails, Piccadilly 6

Williams-Ellis, Clough 66

Working Party of the Design and Layout of Public Offices 100, 103